Medieval Textual Cultures

Judaism, Christianity, and Islam – Tension, Transmission, Transformation

Edited by Patrice Brodeur, Assaad Elias Kattan, and Georges Tamer

Volume 6

Medieval Textual Cultures

Agents of Transmission, Translation and Transformation

Edited by
Faith Wallis and Robert Wisnovsky

DE GRUYTER

The hardcover edition of this book was published in 2016.

ISBN 978-3-11-060138-1
e-ISBN (PDF) 978-3-11-046730-7
e-ISBN (EPUB) 978-3-11-046570-9
ISSN 2196-405X

Library of Congress Cataloging-in-Publication Data
A CIP catalog record for this book has been applied for at the Library of Congress.

Bibliographic information published by the Deutsche Nationalbibliothek
The Deutsche Nationalbibliothek lists this publication in the Deutsche Nationalbibliografie; detailed bibliographic data are available on the Internet at http://dnb.dnb.de.

Printing and binding: CPI books GmbH, Leck
♾ Printed on acid-free paper
Printed in Germany

www.degruyter.com

Acknowledgements

On behalf of the other members of the McGill Research Group “Transmission, Translation and Transformation in Medieval Textual Cultures” – Professors Jamie Fumo, Carlos Fraenkel, Cecily Hilsdale and Jamil Ragep – we would like to thank Québec's Fonds de recherche sur la société et culture (FQRSC) for their financial support during the period when we were an “Équipe en fonctionnement.” Thanks are also due to Dr Jo Ann Levesque, then head of McGill's Office of Research Opportunities, as well as Linda Kemp of McGill's Research Grants Office, for their invaluable assistance as we prepared and submitted our funding application to FQRSC. Dr Edwin (Zoli) Filotas was a great help during the first stage of preparing this volume for review. Dr Walter Young's assistance was especially precious during the formatting and copy-editing of the text; to him we owe a major debt of gratitude.

At De Gruyter, Dr Alissa Jones Nelson expertly shepherded us through the process of review and publication. De Gruyter's anonymous reviewer is also owed our thanks, for the thoughtful and positive comments and suggestions that were made on the first draft. Finally, we would like to thank our contributors for their patience as we gathered and edited the final contributions.

Prof. Faith Wallis
Departments of History and Classical Studies,
and of Social Studies of Medicine
McGill University

Prof. Robert Wisnovsky
Institute of Islamic Studies
McGill University

Table of Contents

Faith Wallis and Robert Wisnovsky
Introduction: Agents of Transmission, Translation and Transformation — 1

Alain Touwaide
Agents and Agencies? The Many Facets of Translation in Byzantine Medicine — 13

Keren Abbou Hershkovits
Galenism at the ʿAbbāsid Court — 39

Charles Burnett and David Juste
A New Catalogue of Medieval Translations into Latin of Texts on Astronomy and Astrology — 63

Warren Zev Harvey
Bernat Metge and Hasdai Crescas: A Conversation — 77

Christine Chism
Transmitting the Astrolabe: Chaucer, Islamic Astronomy, and the Astrolabic Text — 85

Frank T. Coulson
Literary criticism in the Vulgate Commentary on Ovid's *Metamorphoses* — 121

Michael McVaugh
On the Individuality of the Medieval Translator — 133

Raphaela Veit
Charles I of Anjou as Initiator of the *Liber Continens* Translation: Patronage Between Foreign Affairs and Medical Interest — 145

Rosa Comes
The Transmission of Azarquiel's Magic Squares in Latin Europe — 159

A note on the forms of personal and institutional names

The transliteration of Arabic words and names of historic personages follows the norms of the *Encyclopaedia of Islam* 3rd edition (Leiden: Brill, 2007–), save where the reference is to the Latin translation and reception of the author's works (e. g. Rhazes instead of al-Rāzī in the essay by Veit, Azarquiel instead of Abū Isḥāq al-Zarqālluh in the contribution by Comes).

Manuscripts are identified by city, repository, and shelf-mark. The name of the city appears in its English form (e. g. Vienna rather than Wien), but the name of the repository in the language of the country (e. g. Vienna, Österreichische Nationalbibliothek).

The place of publication of printed books appears in its English form (e. g. Naples rather than Napoli).

Faith Wallis and Robert Wisnovsky

Introduction: Agents of Transmission, Translation and Transformation

The McGill University Research Group on Transmission, Translation and Transformation in Medieval Textual Cultures (TTT) is a collaborative enterprise involving historians of medieval culture, literature and art. The working hypothesis of the research group was that medieval textual cultures, regardless of the particular discipline or the specific civilization under review, can best be understood as products of dynamic processes of transmission, translation and transformation. What was transmitted, translated and transformed was the legacy of the civilizations that emerged around the Mediterranean in the last two millennia BCE, particularly in Greece, Rome and the Near East.[1] More specifically, our project stressed the importance of the interchange of ancient legacies for understanding medieval textual cultures (transmission and translation), by recognizing that reception is a creative cultural act (transformation). This approach also evaded any temptation to reduce medieval textual cultures to mere receptacles of the ancient legacies. Placing the accent on the creative aspect of reception avoids two paradigms that in the past have shaped the perception of medieval intellectual history. First, in terms of transmission *across time*, the paradigm of "the survival/revival of the classical tradition" yields to an approach which emphasizes the agency of learned communities in creating and naturalizing this tradition. Secondly, in terms of transmission *between sibling cultures*, we could set aside an instrumentalist view that reduced the historic role of Islam to that of preserving and relaying ancient learning to Europe. Not only is Islam's creative development of ancient concepts better appreciated now, but the complex interaction between these Islamic reformulations and medieval Christian and Jewish textual cultures, each equipped with its own acquired practices of selection and transformation, is changing the way in which this process is understood. We hope to contribute to this new understanding.

The main methodological implication of our approach is that the centrality of transmission, translation and transformation to medieval textual cultures can

1 These ancient civilizations, moreover, were themselves already intertwined as recent scholarship on the dependence of Greco-Roman cultural forms on Near Eastern antecedents has shown: see in particular William W. Hallo, *Origins: the Ancient Near East Background of Some Modern Western Institutions* (Leiden: Brill, 1996), and Walter Burkert, *Babylon, Memphis, Persepolis: Eastern Contexts of Greek Culture* (Cambridge MA: Harvard University Press, 2004).

be fully grasped only within an interdisciplinary framework. This framework challenges linguistic boundaries (Greek, Latin, Hebrew, Arabic, the vernacular languages), religious boundaries (Judaism, Christianity, Islam), geographic boundaries ("East" and "West"), and boundaries of genre (philosophy, literature, science, art). The research group's working assumptions and their methodological implications were subjected to searching and productive discussions at two interdisciplinary workshops on "Vehicles of Transmission, Translation and Transformation" (2007)[2] and "Agents of Transmission, Translation and Transformation" (2010). The present volume is the fruit of the second workshop.

The goal of the 2010 workshop was to build upon the "what" questions addressed in our "Vehicles" workshop of 2007 – specifically, the question of "*What* were the scholarly and literary vehicles by which the shared cultural forms inherited from antiquity were transmitted, translated and transformed?" In the second workshop, we focused on the people, relationships, societies and institutions who were doing the transmitting, translating, and transforming, in an effort to answer the question, "*How* did these agents propel the processes of transmission, translation and transformation?" The transition from vehicles to agents proved to be more complex than we had initially envisioned. The papers presented at the Agents workshop, and the discussions that ensued, made it clear that not every *agent* acted with an *agenda*. And even in those cases where an agenda could be identified, that agenda was sometimes motivated by immediate needs, or by religious and moral factors, that were compelling to the actors, but that are more opaque to us. Some particularly pertinent methodological issues surfaced: for example, what exactly do we mean when we say that a text becomes "available" for transmission or translation? And why do some texts, once transmitted, fail to thrive in their new milieu? In other words, should we expand the notion of agency to account for a decision to *discard* some transmitted texts? In light of our ultimate ambition to re-frame the history of medieval textualities, the second workshop was a turning point, moving us towards a more sophisticated framework of the "ecology" of transmission where not only individuals and teams of individuals, but also social spaces and local cultures, served as agents who shape these textualities.

The TTT team was initially reluctant to adopt the theme of "agents" because it seemed impossible to arrive at a definition of agency that was both clear and specifically applicable to transmission, translation and transformation in medi-

2 The proceedings of this workshop appeared as *Vehicles of Transmission, Translation and Transformation in Medieval Culture*, ed. Robert Wisnovsky, Faith Wallis, Jamie Fumo and Carlos Fraenkel (Turnhout: Brepols, 2011).

eval cultures. In the end, we agreed to give the idea a chance, even without an agreed definition. Our aim was modest: to assemble a provisional tool-kit of concepts and models of agency that would help us to understand something (if not everything) about processes taking place in different cultural settings in the medieval period. From that pragmatic standpoint, etymology, while not as satisfactory as definition, could at least give us a conceptual rope to grasp as we ventured into the unknown.

The Latin verb *agere* that underlies the English words "agency" and "agent" has an intriguing semantic range. It denotes activity, control and purpose; but can also evoke process without a view to an end, or even, oddly enough, passivity. *Agere* suggests action that is direct and even violent: forcing, pushing, throwing, agitating. It is the verb used to convey the action of storms or gusting winds. *Agere* means "to drive (cattle)" but also "to drive off (cattle)," that is, to steal or plunder. The coercive connotation of *agere* extends to provoking or inciting, routing, driving someone mad, condemning someone to exile or to the gallows, or taking someone to court. The element of control moves into the foreground when *agere* means to drive a chariot, ride a horse, steer a ship, lead an army into the field, and manage or administer. From control, *agere* moves easily into the domain of purposefulness: to take action, adopt a policy, perform, achieve, accomplish, or bring something about, to make, construct or produce, to strive for something, to aim at something, even to be "up to something." The imperative *age* is a summons to action or attention ("Move it!" "Look out!"). *Nihil agere* means to be idle, but also to play the fool or to jest – in other words, to be an "actor" in the sense of a stage performer. Indeed, *agere* can mean to "act like" or "behave as" in the positive or neutral sense, but also in the negative sense of pretending. Moreover, the forceful action suggested by *agere* need not be intentional: it can denote the discharge of liquids, the emission of flames by a fire, or the sprouting of roots and shoots. Indeed, when impersonal circumstances are in play (*rebus agentibus*), it usually means that human agency is rendered impotent by *force majeure*. *Agere vitam* means simply "to live one's life," and the verb covers a whole range of meanings involving passing the time, experiencing the change of the seasons, celebrating a feast, staying busy, and just carrying on or proceeding. No end need be in view. Finally, and perhaps most importantly from our perspective, *agere* connotes discourse and relationships between human beings. It signifies, for instance, to have dealings with or to be involved with someone, or to transact business. It also means to discuss, to reason about, to argue, to make a point or state a case. In classical Latin, *agere* usually means "to discuss in writing," but it can also signify "to make a speech." To express thankfulness is *gratias agere*; to voice praise is *laudas agere*.

It is instructive to track comparable or contrasting examples of this semantic cluster in the many languages represented in the medieval worlds covered by our workshop. For example, in Arabic, the term *fiʿl* translates three different Greek terms. The first refers to *action* (i.e., acting-upon, *poiēsis*), which is the opposite of *passion* (being-acted-upon, *paskhein*) (cf. Aristotle's *Categories*). The second refers to an *act* (i.e., deed, *praxis*), which is the product of an agent who possesses intellect and will; an act has moral valuation (cf. Aristotle's *Nicomachean Ethics*). The third refers to *activity* (i.e., the state of actuality, *energeia*), which is the opposite of *potentiality* (i.e., potency, *dunamis*); activity is the product of a cause, which brings something that is in potentiality into activity. What is more, activity is not necessarily a product of intellect, since many changes are natural changes (cf. Aristotle's *Physics* and *Metaphysics*). The point is that the Arabic term *fiʿl* does not refer to the same thing in all the contexts in which it appears. Sometimes it refers to action, other times to act, and other times to activity.

For the moment, however, the Latin root of "agent" and "agency" will simply be conscripted to serve as a provisional starting point for reflecting on the workshop's themes. What we have, then, is a term that expresses "doing" in a particularly purposeful and effective sense, and at the same time encompasses the idea of just letting things happen. It has positive connotations of achievement and direction, and negative connotations as well (plunder, coercion, play-acting...). It is a word about the use of words, about human interactions, about dealing with others and making deals. Let us see where it can take us.

Particularly since the "linguistic turn," historians have been wrestling with the theoretical and methodological complexities of speaking about the motivation of historical actors or assessing the connection between motivation and events. This workshop made us aware that transmission was often the product of an agenda motivated by immediate needs and situations that were evident to the actors, and yet elude us. The first two essays in this collection look at the spectrum of purposefulness – from decentralized and unorganized, through to focused and instrumental.

In "Agents and Agencies? The Many Facets of Translation in Byzantine Medicine," **Alain Touwaide** addresses the re-appropriation by Byzantine translators of ancient Greek botanico-pharmaceutical knowledge through translations of Arabic texts that were themselves grounded in these ancient sources. To date, 120 codices containing around 60 such texts translated between the 10th and 14th centuries have been identified, but these constitute evidence of translation *activity* rather than a translation *program*. The products were practical manuals, and their diffusion was apparently limited. In Constantinople, the figure of Symeon Seth (11th c.) marks the first major breakthrough of Arabic medicine. Symeon is also typical, in that he seems to have worked outside any scholarly group or patronage network. Sicily in the 12th century was the conduit for the Greek translation of *Provisions for the Traveller and the Nourishment of the Settled* by al-Jazzār, known as *Viaticum* in Latin and *Efodia* in Greek. The Greek text presents the remarkable spectacle of a multi-layered composition, with revisions and additions by a number of practitioners – in short, a work of spontaneous, unmanaged cooperation among users of the text. Translation activity in this type of material also varies along genre lines. There are collections of formulae for remedies translated by individuals who traveled from the centre of the Empire into the Islamic lands, particularly Persia. Translations of formal treatises, on the other hand, tend to be made by people hailing from the frontier zone between Byzantium and the Islamic regions. The third group comprises reference materials like Arabic-Greek lexica of plant names. Evidence of collaboration between a native Greek-speaker and a non-speaker can be detected on the pages of a codex written in Constantinople in the 14th century, and suggests a degree of physical mobility between the two cultural spheres. The historical moment may have played a role here, for the end of the caliphate and the recovery of Byzantine control of their capital city may have attracted to Constantinople Arab-Islamic physicians in need of employment. In sum, this translation activity was overwhelmingly practitioner-driven and unorganized. Its extent and diversity, on the other hand, are impressive.

Keren Abbou Hershkovits's paper on "Galenism in the ʿAbbāsid Court" looks at the other end of the purposefulness continuum: a targeted translation program. Abbou Hershkovitz argues that Arab-Islamic learned medicine did not become "Galenic" by virtue of inherent superiority to other kinds of medicine, or even (at first) thanks to court patronage. The early ʿAbbāsid caliphs seem to have been eclectic and opportunistic in their choice of personal physicians, and Galenic doctors were not always their top choice. To gain dominance at court, Christian "Galenic" physicians had to market their product as superior on grounds that re-defined and also transcended their reputation for efficacy. At the same time, they needed to police admission into their own ranks, and hence admission to the court. Textual knowledge of a corpus of Galen's works that had both philosophical interest as well as clinical value became the key to admission into this circle, defining the Galenic doctor as a scholar as well as a practitioner. Hence these works had to be available in Arabic. In sum, the precariousness of the situation of Galenic doctors at the court was a powerful motive for translating Galenic texts and transforming these texts into a sort of curriculum. Many intriguing questions emerge from this analysis. Indian medicine seems to have gone into eclipse after the fall of the Barmakid family, but what would have happened had the Barmakids remained in favour? Could Indian medicine, with its rich textual tradition, have been an *intellectual* rival to Galenism? Or would the Galenists have prevailed because Galenism possessed a textual corpus of a particular kind – one that could be packaged as philosophy as well as medicine?

The next two essays offer different reflections on what one might call the ecology of translation. What factors in the cultural environment cause a translated text to fail to thrive in its host culture? In the Latin West, astronomical texts that we know were eagerly and deliberately sought out, such as the *Almagest* of Ptolemy, were not always widely copied. On the other hand, a meaningful appraisal of what it means for a text to be "available" depends on comprehensive documentation and statistical analysis. This is the goal of **Charles Burnett**'s and **David Juste**'s project to catalogue all Latin translations of Arabic, Greek and Hebrew astronomical and astrological texts, including anonymous and suspected translations. The results to date are intriguing. The translations from the Greek were all of ancient authorities – no late Hellenistic or Byzantine writers. The goal seems to have been to replicate the Alexandrian curriculum, which extended from Euclid to Ptolemy. Yet despite the prestige of this corpus, the number of manuscripts is quite small. The classic canon seems to have been overtaken by a separate Arabic astronomical tradition rooted in Islamic Spain. These works comprised treatises on the astrolabe, tables, and critiques of Ptolemy. They were quickly translated into Latin after their composition, and the new works displaced earlier ones. Tables in particular were rapidly superseded. Furthermore, with the passage of time Latin texts displaced Arabic texts in western universities, and these Latin works exist in hundreds of manuscripts. There are some exceptions, such as al-Farghānī's *Thirty Chapters:* John of Seville's translation survives in 68 manuscripts, Gerard of Cremona's in 48. But Ibn al-Haythgam's *On the Configuration of the World* survives in only one manuscript. The transmission of astrological texts is markedly different. Again, the Greek classics fare poorly (three of the four translations of Ptolemy's *Tetrabiblos* survive only in a single manuscript) while Arabic-Islamic authors do well – only now they hail from the central Islamic lands, not al-Andalus, and with the exception of some Fatimid-period materials, were all writing in the 8/9th centuries. Moreover, the number of manuscripts of these texts augments over time. Ought we to ascribe this to the conservative character of astronomical doctrine, and the lack of necessity to discard outdated data?

Local ecologies of transmission move into the foreground in **Warren Zev Harvey**'s essay, "Bernat Metge and Hasdai Crescas: A Conversation." The environment in this case was the royal court of the kingdom of Aragon at Saragossa. In the decades straddling the turn of the fifteenth century, this court was the professional home of Bernat Metge (ca 1340 – 1413), humanist and author of the foundational work of Catalan literature, *Lo Somni* (*The Dream*), and of Hasdai Crescas (ca 1340 – 1410/11), rabbinical authority, philosopher and author of the

Or Ha-Shem (*The Light of the Lord*). Both were educated in Barcelona, and both were royal advisors. They knew one another and talked to one another: indeed, a record survives of a three-way conversation between Metge, Crescas, and Queen Violant (Yolande de Bar) in 1390, which involved Crescas translating a Hebrew text into Catalan. They also functioned as agents of transmission to each other, and to their respective communities. Metge was a window for Crescas to the Latin classics and contemporary Italian humanism; Crescas was a window for Metge to Jewish traditions, including Jewish interpretations of Aristotle. Traces of their conversation can be detected in their respective works. For example, in *Lo Somni*, Metges presents a chain of Old Testament proof-texts for the immortality of the soul. The ultimate source of some of these texts is Abraham ibn Daud, but the proximate source is Crescas, and they appear in the same sequence in *Or Ha-Shem*. Metge must have acquired them through a conversation with Crescas, since *Lo Somni* antedates *Or Ha-Shem*. This act of transmission is unacknowledged and unobtrusive, but it adds detail and credibility to our picture of the court of King Joan I and Queen Violant as a multi-lingual and interconfessional "ecosystem" that sustained transmission, translation and transformation.

Medieval people's awareness of agency in relation to transmission, translation and transformation is the dominant theme in the papers by Christine Chism and Frank T. Coulson. **Christine Chism**'s "Transmitting the Astrolabe: Chaucer, Islamic Knowledge and the Astrolabic Text" observes that Chaucer's highly popular treatise on the astrolabe contains a history of the transmission of this instrument that is not linear (i.e. not a conventional *translatio studii*) but multi-centric. Knowledge of the astrolabe also leaves considerable agency to the individual, in that the astrolabe is an instrument which has many applications, and which needs to be customized to location. This fluidity and multi-directionality bestows a kind of power on the individual, who is encouraged by Chaucer to perform practical experiments and exercises. At the same time, Chaucer exhibits a certain defensiveness about knowledge of the astrolabe, as if it posed some kind of threat that needed to be neutralized. This awareness of multiple centres of agency in a transmission that is open-ended may arise from the nature of the object itself, whose interchangeable plates reproduced the sky in different latitudes or climates. Thus as one moved from (let us say) England to Rome on a pilgrimage, a passage through different linguistic zones and political jurisdictions would be mirrored by adjustments to the instrument. Can we speak, then, of an agency of the object itself in the transmission and transformation of knowledge about that object?

Reader agency and author agency also underpin **Frank Coulson**'s analysis of "Literary Criticism in the Vulgate Commentary on Ovid's *Metamorphoses*."

The focus here is the implicit awareness of the author of the Vulgate Commentary (who wrote in the region of Tours, *c.* 1250) that an Ovidian tradition was being shaped, and even invented, in his own medieval milieu. Unlike Virgil, Ovid came to the Middle Ages without an authoritative ancient commentary; moreover, he only became a popular poet in the eleventh century. The Vulgate Commentary, while summarizing prior scholarship on the *Metamorphoses*, adopts an open-ended and eclectic approach to the text. At the same time, it shows a marked interest in documenting Ovid's literary influence on authors from the period 1100 – 1250: Alan of Lille, Bernard Silvester, Alexander of Villedieu, Walter of Châtillon, Everard of Béthune, and even the commentator's exact contemporary, Bartholomaeus Anglicus. The Vulgate Commentary in effect showcases writers from his own age and milieu as the primary agents of transmission and transformation of Ovid.

The concept of agency embedded in the word's etymology seems inherently biased towards the individual, the "actor" who drives the chariot, leads the army into battle, formulates a purpose and brings something about. While our TTT project concerns itself with explaining processes that transpire over centuries and across cultures, and so is perhaps unconsciously biased against the individual, this workshop forced us to pay attention to ways in which thinking about the contingent and the individual can enrich synthetic analysis. In particular, **Michael McVaugh**'s "On the Individuality of Translators" is a salutary reminder that a translator constantly exercises agency by deciding what word he will choose to represent the term in his source. These choices can open larger historical vistas, though the prospect is not always clearly visible, even when we have several translations from a single pen. For example, Arnau de Vilanova's translations of two medical works from the Arabic vary widely in quality: was the cruder translation the earlier work, and the more polished one a product of his maturity? Or does the difference reflect conditions of production (the cruder version being a working draft made for Arnau's own use)? Agency can be brought into somewhat sharper focus when we compare two translations of the same work, where the translators diverge in their choice of strategies. The Latin translations of Maimonides's *On Asthma* are a case in point. The version by the Montpellier master (and nephew of Arnau de Vilanova) Armengaud Blaise is longer by 10 % than the one by the Italian Giovanni da Capua. It is also more polished in style, more academic in diction and presentation, and marked by unconscious echoes of the New Testament – all markers of Armengaud's and Giovanni's different professional and confessional profiles. There are two Catalan translations, made almost at the same time, of the *Surgery* of Teodorico Borgognoni: one by a practitioner, Guillem Corretger, and the other by the academically trained Bernat de Berriac. Both translations are quite faithful and literal. Bernat used Guillem's translation of books 1–3, but made a new version of book 4. However, a manuscript now in Graz shows Bernat's own correction to Guillem's translations of books 1–3. In listing recipe ingredients, Bernat restores the technical term *ana* ("each" – to indicate that the same quantity of all the ingredients in the list is to be employed) where Guillem used a circumlocution. Other individuals are engaged in the work of transmission in this Graz volume, notably a Jewish reader (a practitioner in Majorca?) who rendered the Catalan chapter titles into Hebrew in the margins. The Jewish hand also added recipes translated from the surgery textbook of Lanfranc of Milan. But was he translating from the Latin Lanfranc or the Catalan Lanfranc? In sum, the finger of the individual points to a landscape whose contours beckon, and simultaneously fade from view.

It is important to recall that when medieval sources name the *actor/auctor* of a cultural production – especially, but not exclusively, a work of art or architecture – they are not necessarily referring to the artist or the architect, but rather to the patron or *Bauherr* ("building-lord" – the abbot, bishop or prince who had the authority to dictate the form of the edifice). Perhaps because our own professional sympathies lie with the scholar and translator, we must constantly be reminded of the agency of kings and rulers. **Raphaela Veit**'s essay on "Charles I of Anjou as Initiator of the *Liber Continens* Translation: Patronage between Foreign Affairs and Medical Interests" tracks the agency of this king in locating a manuscript of the *Kitāb al-Ḥāwī* of Rhazes (al-Rāzī), commissioning its translation, and designing its manuscript presentation. The prologue attached to the translation tells how Charles, whose realm (at least notionally) encompassed Anjou, Provence, Jerusalem, Achaia and the Kingdom of Sicily, sent a delegation to the "king of Tunisia" to ask for a copy of the *Liber continens*. He then hired as translator "a reliable man, who had mastered the Arabic and the Latin language," identified in the explicit as Master Faragius (Faraj ibn Sālim), a Jew of Agrigentum. Official documents from Charles's chancery confirm that the translation was begun on 6 February 1278 and completed 12 February 1279. The archives also contain accounts of the payments to the translator, and interim reports on his progress. They tell us that the Arabic original was in five volumes, that it was in the custody of the royal treasurer, and that only one volume at a time was released to the translator. The treasurer also arranged for the packing and transportation of the completed translation. Once he had received Feragius's text, Charles submitted it for "peer review" to physicians at his court and at the universities of Naples and Salerno, who were unanimous in their approval. Charles then commissioned a richly decorated manuscript of the work, now Paris, Bibliothèque nationale de France lat. 6912. The direction of the project was confided to Jean de Nesle, physician and royal librarian, and the miniatures were by a famous illuminator, Giovanni of Monte Cassino.

In his capacity as ruler of Achaia, Charles commissioned a second copy through his chancery in Morea, now probably Vatican City, Biblioteca Apostolica Vaticana Vat. lat. 2398 and 2399. The prologue that chronicles Charles's agency in such detail was probably written by a courtier with academic training in medicine. In stately Latin, it presents Charles as a ruler who not only loves science, but translates his love into action, and is ready to expend effort and money to do so. But why choose Rhazes's *Kitāb al-Ḥāwī*, a sort of miscellany of medical observations and quotations which was neither successful in the Arabic world, nor (as it turned out) destined for great success in the Latin West? Were all the really useful Arabic medical encyclopedias translated already? It would seem that unlike Spain, Sicily contained few Arabic manuscripts for Christian rulers to exploit. Secondly, Charles was pursuing an ambitious Mediterranean foreign policy. Of particular importance was his relationship with Tunisia, where partisans of Charles's predecessors on the throne of Sicily, the Hohenstaufen dynasty, had taken refuge. The transmission of the Rhazes manuscript may have taken place in the context of the diplomatic negotiations between Charles and the emir of Tunis that resulted in the expulsion of these partisans from the emir's lands. It is unlikely that Charles initiated a quest for this particular work; but once in his hands, he was ready to get maximum propaganda mileage out of it. In sum, what we have here is an incident of opportunistic transmission and translation.

The agency of rulers also features in **Rosa Comes's** "The Transmission of Azarquiel's Magic Squares in Latin Europe" – an instance of the importance of magic and other "power sciences" in medieval translation programs. The case in point is the translation by Azarquiel (Abū Isḥāq al-Zarqālluh) of a description of a talismanic magic square in the Castilian *Libro de astromagia* commissioned by King Alfonso X, one of many works on applied astronomy sponsored by that monarch. A magic square is a square figure divided into an equal number of rows and columns to create cells; the cells are then filled with numbers chosen and arranged so that the sum of each row, column and major diagonal is the same. Magic squares are discussed in mathematical treatises, in magic handbooks (where seven types of magic square are connected to the seven planets, and are used for talismanic purposes) and in works which combine both genres. The earliest are in Arabic, from the 10th century, and are purely mathematical. But the first works to be transmitted to Europe were about talismanic magic squares, and their debut appearance was in Alfonso's *Libro de astromagia*. Alfonso commissioned translations of Azarquiel's treatises on astronomical instruments, and it was his admiration for this author that apparently led him to the *Kitāb tadbīrāt al-kawākib* (*Book on the influences of the planets*). However, only one of Azarquiel's planetary squares – the one associated with Mars – is to be found in Alfonso's fragmentary treatise. The chain of transmission then goes underground, only to re-emerge in 16th and 17th century works of occult philosophy by Cornelius Agrippa, Girolamo Cardano, and Athanasius Kircher. However, a Mars magic square that differs somewhat in content (though not construction) from the Azarquiel/Alfonso model survives in several manuscripts from the $13-15^{th}$ centuries. The instructions for constructing a talisman based on the square are roughly the same in both versions. However, Alfonso's version made some intriguing editorial changes to the text that are only found in this version. One of the "powerful men" who could be subdued by the square is a bishop (Alfonso's relationship with the ecclesiastical hierarchy was notoriously tense). Like Charles of Anjou, Alfonso apparently conceived of his cultural projects as part of a broadly political agenda.

Our collection closes with a reflection on the roots of the TTT concept and method in the scholarship of the 20th century. **Carlos Fraenkel** publishes the torso of a research project sketched in 1959 by Shlomo Pines, the eminent historian of philosophy and religion, as part of an application for research funding. Rather provocatively, Pines proposed embedding the history of medieval Jewish philosophy within the history of medieval Arabic thought. It is hard to underestimate the implications of such an approach: what would it mean to think of the sources of Jewish piety, theology and philosophy as springing from Arabic fountainheads? Fraenkel argues that Pines's embryonic project remains a desideratum. A case in point is Maimonides, who felt completely at home in the milieu of Arabic philosophy and its particular approaches to the ancient Greek legacy. Even the *Guide for the Perplexed* is essentially a work of "philosophical religion" deeply rooted in the achievements of Maimonides's Arab-Islamic intellectual milieu. Is this sufficient to justify Pines's stronger thesis that Jewish philosophy is a branch of the tree of Arabic thought? Perhaps not, but the fact that a scholar of Pines's wide-ranging and deep erudition was ready to try out this idea speaks to the robustness and appeal of the model of dynamic translation, transmission and transformation that we seek, in this volume, to expose.

Alain Touwaide

Agents and Agencies? The Many Facets of Translation in Byzantine Medicine

1 Introduction

Many historical inquiries have been devoted to translation in Byzantium and the Byzantine World. Most have dealt with the rendering of the classical legacy from Greek into such languages as Latin in the West, and Syriac, Georgian, Armenian, Coptic, and Arabic in the East.[1] Studies are usually framed in research programs on the diffusion and circulation of classical knowledge from Byzantium to its periphery, and have two possible aims: using such versions as further witnesses to texts to be critically edited, or tracing texts whose Greek version is no longer available. Only in rare cases, until recently, have they investigated translations from a language of the neighbouring cultures of Byzantium into Greek.[2]

In this contribution I will focus on a set of medico-pharmaceutical texts to be found in Byzantine manuscripts that are either explicitly identified or can be identified as translations of treatises or fragments of treatises by Arabic scientists, or of information coming from populations identified in the texts as "Syrians," "Persians," or "Saracens." In the context of my research on the transmission of botanico-pharmaceutical knowledge around the Mediterranean from Antiquity to the Renaissance and the dawn of modern science, passing through Byzantium, the Arabic world, the West,[3] knowing who the translators were, in

1 For an overview, see Touwaide, Transfer of Knowledge, pp. 1368–99.
2 For an example of such a "reverse" process of translation, see Fisher, Greek Translations of Latin Literature, pp. 173–216.
3 For a general presentation on each of these topics, see the following (by topic, in chronological order of publication): Touwaide, Pharmakologie, cols. 215–222. On Antiquity: *idem*, Healers and Physicians, pp. 155–73. On Antiquity and Byzantium: *idem*, Strategie terapeutiche, pp. 353–73. On the Arabic World: *idem*, Transcultural Tradition, pp. 175–78; Persistance de l'hellénisme, pp. 49–74; Theoretical Concepts, pp. 21–39; Traduction arabe du *Traité de matière médicale*, pp. 16–41; Tradition and innovation, pp. 203–13; Paradigme culturel et épistémologique, pp. 247–73; and Intégration de la pharmacologie grecque, pp. 259–89. On Antiquity, Byzantium and the West: *idem*, Legacy of Classical Antiquity, pp. 15–28. On the West: *idem*, Fuentes de la terapia medieval, pp. 29–41; Enfermedad y curación, pp. 155–66; Pharmacology, pp. 394–97; Pharmacy and materia medica, pp. 397–99; and Pharmacology, Pharmacy, pp. 1056–1090. On the Renaissance: *idem*, Botany and Humanism, pp. 33–61; *Ancient Botany*; and *Loquantur ipsi ut velint*, pp. 151–73.

what circumstances they worked, for whom and for what purposes they carried out their activity, is of fundamental importance.[4]

2 Status quaestionis

Until not so long ago, only some publications had been devoted to this topic. Most of them focused on a work that I shall examine in more detail: the Byzantine translation of the *Zād al-musāfir* by the Arabic physician Ibn al-Jazzār (898–980 CE),[5] known in Byzantine studies under its Greek title *Efodia tou apodêmountos.*[6] A fragment was published in the late eighteenth century by the Dutch physician and philologist Johannes Stephan Bernard (1718–1793).[7] Although Bernard realized that the text he published is similar to the *Viaticum peregrinantium* by Constantine the African (died after 1087),[8] that is, the Latin version of the *Zād al-musāfir,* he did not understand the exact nature of the Greek work; he wrongly attributed it to a certain Synesius (who could not, however, be the famous Synesius of Cyrene [ca. 370–ca. 413 CE],[9] for the chronological reasons Bernard rightly advocated) on the basis of incorrect information in the manuscript in which he found the text.[10] Many other scholars approached the *Efodia* after Bernard, including the French historian of medicine Charles Daremberg (1817–1872),[11] the Vatican *scriptor* Giovanni Mercati (1866–1957),[12] and the

4 On translators' methods, see, for instance, the essays in Hamesse, ed., *Les traducteurs au travail.*

5 On Ibn al-Jazzār, see Ullmann, *Die Medizin im Islam*, pp. 147–49; and Sezgin, *Geschichte*, 3: 304–7. See also Micheau, Connaissance d'ibn al-Jazzâr, pp. 385–405; and Ammar, *Ibn Al Jazzar & the Medical School of Kairouan.* For an edition of Book 6, see Bos (transl.), *Ibn al-Jazzār on sexual diseases.*

6 For a brief note about this work, see Hunger, *Hochsprachliche profane Literatur*, pp. 306–307.

7 Johannes Stephanues Bernard, *Synesius de febribus, Quem nunc primum ex codice MS. Bibliothecae Lugduno Batavae edidit, vertit, notisque illustravit.Accedit Viaticum Constantino Africano interprete lib. VII.* Pars. Amstelaedami: Apud Gerard de Groot, et Lugduni Batavorum: Apud Philippum Bonk, 1749.

8 On Constantine the African, see, most recently, Green, Constantine, pp. 145–47, where the author suggests that Constantine was not from Carthage as usually stated, but from Cairouan.

9 On Synesius, see Baldwin, Synesius of Cyrene, p. 1993.

10 This manuscript was Leiden, Bibliotheek der Rijksuniversiteit, *Vossianus graecus* F 65, which was already in Leiden in Bernard's time.

11 Daremberg, *Notices et extraits.* See, particularly, pp. 63–100 of the 1854 edition, the section entitled "Recherches sur un ouvrage qui a pour titre *Zad el-Mouçafir*, en arabe, *Ephode*, en grec, *Viatique*, en latin, et qui est attribué, dans les textes arabes et grecs, à Abou Djafar, et, dans le texte latin, à Constantin."

Italian Arabist Francesco Gabrieli (1904–1996),[13] to mention just a few. But neither they, nor any subsequent classical philologists or Byzantinists undertook a specific study of the treatise, which is still unpublished in spite of repeated announcements of a critical edition by the Greek philologist and historian of medicine Gerasimos Pentogalos.[14]

Although the Greek historian of medicine Aristoteles Kouzis (1872–1961) published, in 1939, an article entitled "Quelques considérations sur les traductions en grec des oeuvres médicales orientales et principalement sur les deux manuscrits de la traduction d'un traité persan par Constantin Méliténiotis" ("Some Considerations on the Translations into Greek of Oriental Medical Works and Principally on Two Manuscripts of the Translation of a Persian Treatise by Constantin Meliteniotis"),[15] the works he studied did not receive much attention in the subsequent decades. Instead, interest in translations from oriental languages into Greek seems to have come more from the history of astronomy. As early as 1952, Joseph Mogenet (1913–1980) discovered the most ancient trace of Arabic astronomy in Byzantium in the marginal notes of a Vatican manuscript.[16] He opened a new field of investigation that was further illustrated by his follower and successor, Anne Tihon,[17] and also by other historians of medicine, science or Byzantine literature.[18]

3 Preparatory Research

In the footsteps of Joseph Mogenet, under whose direction I had the privilege of beginning to undertake work on my PhD thesis,[19] I started investigating the translations I shall examine here, even though I did not publish on the topic be-

12 Mercati, Filippo Xeros reggino, pp. 9–17.

13 Gabrieli, Zâd al Musâfir, pp. 205–20.

14 See Duffy (ed. and transl.), *Ioannis Alexandrini*, p. 20, n. 1.

15 Kouzis, Quelques considérations, pp. 205–20. On Meliteniôtês see Hunger, *Hochsprachliche profane Literatur*, pp. 313–34, n. 6; Eftychiadou, *Eisagôgê*, p. 310.

16 Mogenet, Une scolie inédite, pp. 198–221; and subsequently: *idem*, L'influence de l'astronomie arabe, pp. 44–55.

17 See Tihon, Tables islamiques, pp. 401–25; and *eadem*, Les tables astronomiques persanes, pp. 603–24. More recently, see *eadem*, Les textes astronomiques arabes, pp. 313–24.

18 See, for example, Harig, Von den arabischen Quellen, pp. 248–268; Congourdeau, Le monde byzantin, pp. 271–73; *eadem*, A propos d'un chapitre, pp. 261–77; or Mavroudi, Exchanges, pp. 62–75.

19 Touwaide, Les deux traités de toxicologie attribués à Dioscoride.

fore 1991.[20] Although I capitalized on the few available works in the history of medicine, I began with the basics, that is, locating the manuscripts containing such texts. In 1997 I systematically browsed available printed catalogues of Greek manuscripts,[21] trying to trace any text explicitly presented, in its title, as a translation of Arabic or Persian treatises, or containing elements that suggested such an origin, or, in a broader sense, a contact with populations once in the Arabic empire.[22] As a result of this preliminary inquiry, I identified more than 120 codices in collections worldwide containing *prima facie* some 60 texts.[23] On this basis, I traveled to personally inspect these manuscripts, to analyze them from a codicological and paleographical viewpoint, and, from this, to possibly reconstruct their history.

Ten years later, in 2000 – 2001, I had an opportunity to devote myself more exclusively to this topic in the context of a programme on "the scientific interrelations between Europe and Islam during the period 1300 – 1800 and/or with comparisons between their respective scientific traditions," directed at that time by Professor Jamil Ragep at the University of Oklahoma. With an entire year to decipher, from manuscripts, the texts I had identified earlier, I was able to transcribe the full text of at least one version of each of the treatises I was interested in, including the *Efodia* above. The versions I transcribed were identified on the basis of my codicological analyses of the manuscripts – mainly their date or period of production and their place of origin – and a preliminary (actually macrostructural) analysis of the texts in each codex, so as to determine what might possibly be the most ancient and most complete versions of the works under consideration. For several texts, I was able to collate more than one copy and to draft a *stemma codicum*.

---

20 Touwaide, Un manuscrit athonite, pp. 122 – 27.

21 I took as a starting point the classical catalogue edited by Hermann Diels, *Die Handschriften der antiken Ärzte* (and supplement). However useful it has been, my further work has shown that Diels's catalogue is not as accurate and complete as one would wish, particularly for the texts whose manuscripts I was trying to locate. See Touwaide, Greek Medical Manuscripts, pp. 199 – 208; and *idem*, Byzantine Medical Manuscripts, pp. 453 – 595. I also took advantage of the following works: Sinkewicz, *Classical and Late Antiquity*; *idem*, *Patristic and Byzantine Periods*; and Sinkewicz and Hayes, *Palaeologan Period*. Nevertheless, it was indispensable to check all catalogues of Greek manuscripts, taking as a basis Olivier, *Répertoire*.

22 In spite of the differentiated origin of the texts under consideration here – both geographically (mainly East of the Byzantine empire, but also including Northern Africa) and linguistically (Arabic, but also Persian) – I will, as a convention designed to ease communication, identify this entire production as *Arabic*.

23 Touwaide, *Medicinalia Arabo-Byzantina*.

Although I have not yet published the results of my work – they are too many to be easily and quickly released – I have nevertheless published a preliminary report[24] and data on some specific aspects of these translations.[25] More recently, some publications by other authors have been devoted to such translations, possibly taking advantage of my own work.[26]

4 Toward an Analysis

My transcription of texts quickly made it clear that the textual body I was interested in is artificial, as it is made of very different elements, not only in terms of period and place of origin, but also with regard to the translators, the material they translated, the methods and circumstances of their work, and the reasons why they translated. A focus on translation as a process soon revealed that the very notion of translators as specialized professionals, possibly working in an organized context (perhaps for a wealthy patron who commissioned the translation), was not necessarily applicable.

During this period of work, the translation process also became increasingly the object of specific studies, some of which led to a dramatic revision of classical theories. It will suffice to mention here the investigations by Marie Geneviève Guesdon on the *Bayt al-Ḥikma*[27] and the research by Vivian Nutton on Jundishapur.[28] This does not mean that translation agencies did not exist, but invites us to be extremely cautious in hypothesizing agencies and/or groups of translators working in collaboration, or in coordination, when analyzing treatises presented as translations, or clearly resulting from a translation, and possibly presenting some affinities of whatever nature.

The documentation under scrutiny – that is, both the texts and their manuscripts – is informative from this viewpoint. It is relatively late, and also pretty limited in time, as it started at the earliest in the tenth century and did not ex-

---

24 Touwaide, Arabic Medicine, pp. 45 – 53.

25 See, in chronological order of publication: Touwaide, Arabic Materia Medica, pp. 223 – 46; Arabic Urology, pp. 167 – 73; *Magna Graecia*, pp. 85 – 101; Medicina Bizantina e Araba, pp. 39 – 55; The Jujube Tree, pp. 72 – 100; Kidney Dysfunction, pp. 12 – 20; and Arabic into Greek, pp. 195 – 222.

26 See the following works by Mario Lamagna: La *recensio amplior* inedita, pp. 271 – 80; La *recensio amplior*, pp. 321 – 33; Per l'edizione del *De urinis*, pp. 15 – 26.

27 See Balty-Guesdon, Le 'Bayt al-Hikma' de Baghdad, pp. 131 – 50; *Bayt al-Hikma* et politique culturelle, pp. 275 – 91.

28 Nutton, Jundîshâbûr, p. 22.

tend beyond the fourteenth. For several texts – particularly those of the later part of the period – we have the original of the translation or at least a copy very close to it. Furthermore, many of the texts were practical manuals, not artistic or deluxe productions aimed at display, gift, or any of the other socio-economic functions that books can sometimes have. Finally, although there was no clash of civilizations – though some indeed have been tempted to imagine otherwise[29] – this production was limited, and, apart from one remarkable exception (the *Efodia*), did not circulate widely – even though it was further transmitted from Byzantium to the West during the Renaissance.[30]

A combination of philological and paleographico-codicological approaches makes it possible to distinguish two major loci for the translation of Arabic medical treatises into Greek: Constantinople and Sicily, with several periods for the Constantinopolitan activity. To each of these places and/or periods corresponds a different type of translation, that is, a different profile of translator, a different rationale for translation, a different function, and also a different diffusion.

Probably the most ancient text in this study is presented as a letter by an Arabic physician, accompanying a medicine offered to the Byzantine emperor.[31] The text is brief and does not contain many explicit chronological elements. The codicological analysis is not of great help as the only manuscript in which it appears is posterior.[32] However, from the name of the emperor[33] we can hypothesize a tenth-century work. If the explicit elements of the title are to be accepted, this is not properly a translation but rather a process of cross-cultural diffusion, initiated by an Arabic physician who was possibly bilingual, or who was associated with a bilingual translator. This physician probably wished to illustrate the achievements of Arabic medicine following the assimilation of the Greek legacy

29 See Huntington, *Clash of Civilizations,* with its epiphenomenon in Lewis, Roots of Muslim Rage, pp. 47–60.

30 See, for example, Rāzī's treatise on smallpox. A Latin translation of its Byzantine Greek translation was published as early as 1498 by the Italian humanist Giorgio Valla (1447–1500): Valla, *Giorgio Valla Placentino interprete.* The Greek text was published fifty years later by the French humanist-physician Jacques Goupyl (d. ca. 1564): Goupyl, *Alexandri Tralliani Medici.* On this volume, see Förstel, Alexandre de Tralles, pp. 27–30.

31 See the text by a certain *Filius Ameroumanae Afri* (*uios tou amerourmnê afrikês*), still in manuscript at Oxford (*Codex Baroccianus* 150), further on which may be found in Coxe, *Catalogi Codicum*, col. 262.

32 Fifteenth century, according to Coxe, *Catalogi Codicum.*

33 In the text, the emperor is identified as *Rômanos Porphyrogennêtos*, who could possibly be Romanos II Porphyrogenitus (regn. 959–963 C.E.). On him, see Kazhdan, Romanos II, pp. 1806–7.

through the translation process in the ninth century, if not earlier.[34] If this is correct, this process of transfer came from a practitioner and had some practical motivation, whether medical *stricto sensu* or somewhat propagandistic. It may be correct to consider this text as a display, particularly because the medicine it describes is a compound one, being what we call now a broad-spectrum therapeutic agent. This type of medicine had been particularly developed in Arabic pharmaco-therapy.[35]

In chronological order, the next translation phase takes place in the eleventh century, possibly in Constantinople, and may have had Symeon Seth as its major protagonist. This personage is a complex figure whose facets have come to light only in recent decades.[36] Until recently, he was mainly known as the author of a manual on the dietetic and medical properties of food inspired by Arabic sources,[37] a treatise on natural philosophy based on Aristotle,[38] and a refutation of Galen.[39] More recent studies have been devoted to his Greek translation of the famous tales of *Kalīla wa Dimna*,[40] and, fifteen years ago, an essay proposed him as author of the translation of Rāzī's treatise *On smallpox and measles*.[41] He is also credited with the study of Arabic astronomy.[42]

---

34 Of the abundant literature on the transfer of Greek science to the Arabic world, it will suffice to mention here the recent works by Dimitri Gutas (*Greek Thought, Arabic Culture*); and Peter E. Pormann and Emilie Savage-Smith (*Medieval Islamic Medicine*).

35 See, most recently, Touwaide, Pharmacology, Pharmacy, pp. 1056–1090 (esp. at pp. 1072–76).

36 On Seth, see Brunet, *Symeon Seth,* pp. 13–39; Sjöberg, *Stephanites und Ichnelates,* pp. 87–99; and Kazhdan, Seth, Symeon, pp. 1882–83. Subsequently, see Leven, Seth, Symeon, pp. 330–31; Condylis-Bassoukos, *Stéphanitès kai Ichnélatès,* pp. XXIII–XXV, and, more recently, Bouras-Vallianatos, Galen's Reception, pp. 436–442.

37 Langkavel, ed., *Simeoni Sethii.*

38 The *Conspectus rerum naturalium.* For the Greek text, see Delatte, ed., *Anecdota atheniensia,* pp. 17–89.

39 See Schmid, Eine *Galen-Kontroverse*, pp. 491–5 and, for a recent edition of the Greek text, together with an English translation and a commentary, Bouras-Vallianatos, Galen's Reception, pp. 458–463 (Greek text), 464–469 (English translation) and 442–455 (commentary).

40 On this translation, see the following by Helena Condylis-Bassoukos (in chronological order of publication): À propos de la traduction grecque, pp. 207–12; Stéphanitès kai Ichnélatès, pp. 139–49; *Stéphanitès kai Ichnélatès,* pp. XXIII–XXV; and *Kalīla wa-Dimna*, pp. 41–50.

41 See Congourdeau, Le traducteur grec, pp. 99–111. On Rāzī, see Richter-Bernburg, Razi, pp. 671–74; and Izkandar, Razi, pp. 155–56, with supplementary information in Druart, Razi, pp. 434–6. For the Arabic text of this treatise on smallpox, see Channing, *Rhazes de Variolis et Morbilis*; and, for an English translation, see Greenhill, *Treatise.* For further on his smallpox treatise, see Leven, Zur Kenntnis, pp. 341–54.

42 See Delatte, ed., *Anecdota atheniensia*, pp. 91–136.

The attribution of Rāzī's translation to Seth is not necessarily correct, as it is not supported by any evidence. It was proposed almost by default because the textual tradition of the text goes back to the eleventh century, and Symeon Seth was considered the only possible candidate. Whatever the case, this Greek version of Rāzī's treatise suggests a greater interest in Arabic medicine in Constantinople and the presence of such medicine among scholars and their literary works.[43]

We have no explicit evidence for a group of translators in the Byzantine empire – in Constantinople or any other city – nor for translations of works from the Arabo-Islamic scientific tradition commissioned by any authority, whatever its nature: civil, religious, or otherwise. The key element here is the translator as an individual – whoever he was, Symeon Seth or another individual whose exact identity we do not know – with linguistic skills and an interest in the most recent developments of a neighbouring culture. Unfortunately, we do not have much information about the biography of Symeon Seth, or any other possible translator with a similar profile, at that time in Constantinople. According to the titles of his manuscript works, Seth was originally from Antioch, had a position at the imperial court in Constantinople, and dedicated his *Treatise on the properties of foodstuffs* to Michael VII Doukas (r. 1071–1078). A polymath like his better-known contemporary Michael Psellos (ca. 1017–1078 or later),[44] Seth was more probably a native of an area of multicultural contacts, who moved to the capital of the empire (possibly thanks to his intellectual skills), and succeeded in approaching the milieu of the court. One may hypothesize that, at that time, Arabic science, literature, and, generally speaking, culture, was the latest and most advanced *nouveauté* in Byzantium, even though the most brilliant intellectual in Seth's time, Michael Psellos, was strictly and only Byzantine.

In twelfth-century Southern Italy, we have a very different situation with the *Efodia*. Although, as I have mentioned, the work is still unpublished, it has been the object of much attention because of its most ancient extant manuscript, the *Vaticanus graecus* 300.[45] After several divergent dates or periods were proposed,[46] the codex has been dated to the mid-twelfth century,[47] at a place

---

43 On Arabic medicine in eleventh-century Constantinople, see Touwaide, Un manuscrit athonite, pp. 122–27.

44 On Psellos, see Kazhdan, Psellos, Michael, 1754–1755.

45 On this manuscript, see Mercati and Franchi de' Cavalieri, *Bybliothecae Apostolicae*, pp. 430–37.

46 Among others: the eleventh or twelfth century, according to Mercati and Franchi de' Cavalieri, ibid.; and the second half of the twelfth century, according to Canart, Le livre grec, pp. 103–162 (see, esp., p. 146).

that is still an object of debate.[48] Whatever the actual place, be it the court of Palermo, Reggio di Calabria, or the San Salvatore monastery of Messina, the important point is that it was produced in a formal milieu. This contrasts with the translation itself, which bears abundant traces of circulation and use among practitioners, as I have shown in previous publications.[49] It will suffice to mention here that, already in the *Vaticanus* codex – which is the most ancient of the translations currently known – the text of the *Efodia* is a multi-layered compound, made of accretions of different origins and including a wide range of data, from popular names of plants to references to Hippocrates, Galen, and other learned physicians of classical antiquity.

On this basis, I have suggested that we take the translation of the *Efodia* not to be the result of a commission by an authority carried out in a learned milieu, but rather the result of a natural, almost spontaneous collaboration between the different linguistic and ethnic groups of physicians practicing in Sicily, or on the mainland. In its original, almost bilingual text, technical terms were not necessarily translated into Greek but transliterated from the Arabic into the Greek alphabet. This first version was further improved by introducing, probably in the margins at first, the exact translation of the terms left in Arabic, or other explanations of any kind aimed at clarifying the Arabic text and its references to geography, the calendar, and other elements typical of Arabo-Islamic society and culture. Fortunately, such annotations have not superseded the original text, but have been added to it, so that we can see different layers of data gradually introduced into the text, sometimes one after another. The authors of these notes were not necessarily scholars who sought to improve the text from a literary viewpoint, but most probably physicians who may have used the text in their practice. They did not limit themselves to connecting the text with the reality they were used to dealing with – giving, for example, the exact Greek names of the plants which had been left in Arabic – but they also introduced data from their own practice of medicine into the text.

On the basis of a close analysis of the original text and this set of differently-natured annotations, I have suggested that the translation was made, among practitioners, by physicians of the two cultures in contact with each other and

47 Wilson, The Madrid Skylitzes, pp. 209 – 19.

48 Reggio Calabria, in Canart, *op. cit.*; Palermo, at the Norman court, in Wilson, *op. cit.*, (and also in Cavallo, La trasmissione scritta, pp. 157 – 245 [particularly at p. 197], and *idem*, *La cultura italo-greca*, pp. 495 – 612 [esp. at p. 559]); and Messina, in Re, *A proposito*, pp. 329 – 341; Foti, *Il Monastero*; and Lucà, I Normanni, pp. 36 – 63.

49 See Touwaide, *Magna Graecia*, pp. 85 – 101; and *idem*, *Medicina Bizantina e Araba*, pp. 39 – 55.

working in collaboration. This first-hand text was then circulated among practitioners and revised in a cumulative way, before eventually reaching learned milieux in Sicily or on the mainland. There, it was fixed in a canonical form. This is the codex *Vaticanus graecus* 300, which represents not the origin of the process of diffusion of Arabic science among the Greek populations of Southern Italy, but rather the coagulation of a movement started earlier among practitioners coexisting in Southern Italy. The end result arrived later in an institutional milieu, whether on the island or on the mainland, and whether religious (the San Salvatore monastery) or civil (the court of Palermo). I thus propose this translation be considered a natural expression of the coexistence of different groups interacting in the practice of medicine. In this view, it is probably impossible to ascribe it to a well-defined milieu or to identified individuals, as it is likely to have been made by a group of professionals of the two cultures in contact. In a following stage, the text was received among the literati in an ascending movement from practice to learning, contrary to the movement generally attributed by historians to the production of new works, from the learned spheres of society to the world of practice.

Interestingly enough, once the text of the *Efodia* had been absorbed into learned culture and textually fixed, it circulated from the periphery to the capital of the Byzantine empire in a centripetal movement, opposite to the centrifugal (center-to-periphery) direction all too often supposed by modern scholars to be the usual mode of cultural diffusion. It may be significant that this centripetal movement took place in the late-thirteenth and early-fourteenth century, where we encounter the latest and most important group of translations under consideration here.

The set of translations dating to this period is composed of very different works,[50] from monographic treatises,[51] to shorter notes (e.g., on diagnosis by means of urine analysis),[52] to bilingual lexica of plant names.[53] They also have very different redactional histories, from just one copy[54] to several,[55] from a

50 For all of these works, see Touwaide, *Medicinalia Arabo–Byzantina*. Identification of texts is according to their Greek title translated into Latin, as per traditional philological usage, and for the sake of clarity.

51 See (in manuscript) ʿAbd Allāh, *Tractatus medicae artis*, ff. 1 recto–81 recto.

52 Besides *De urinis* by Avicenna (supra, n. 26), see, for example, the following: Ideler, ed., *De urinis e libro syriaco*, pp. 303–4; *idem*, ed., *De urinis ex Persarum disciplina*, pp. 305–306; and (in manuscript) Anon., *De urinis e Persarum scriptis*, ff. 41 recto–43 verso.

53 For an inventory of such lexica with their manuscripts and available editions, see Touwaide, *Lexica medico-botanica byzantina*, pp. 211–28.

54 See, for example (in manuscript): Syrus, *Medicamentum hepaticum*, f. 353 verso.; Ataamatoulmardaou, *Dieta infirmorum*, ff. 186 recto–206 verso; or Israelitus, *De elaterio*, f. 296 recto.

unique version[56] to as many revisions as existing copies.[57] With regard to the translators, they can be divided into three major categories, as far as I can tell at the current state of research.

The first category includes collections of formulae for remedies; the titles of such collections state explicitly that they have been translated by individuals who have traveled to a land identified as Persia.[58] This recalls the case of astronomy, where we have, indeed, similar texts produced by individuals who knew about the achievements of their colleagues in the former Arabic empire;[59] these traveled to Persia and learned astronomy at local schools, then translated the most advanced treatises and brought them back to Constantinople.[60] Again, we have no knowledge that such individuals operated in a formal context, as part of any organized group, or in translation agencies devoted *in toto* or *in parte* to the assimilation of external science. Nevertheless, we should avoid the fallacies of individualism and autobiographical claims that may have inflated actual facts. Whatever the case, we still need to understand how these expeditions – if we can refer to them as such – became possible in the late thirteenth and early fourteenth centuries; and, as we shall see, the third group of translations under examination for this period will help us on this point.

---

55 The *De alimentorum facultatibus* by Symeon Seth can be read, according to the current state of knowledge, in sixty manuscripts, the *Efodia* in more than forty currently known manuscripts, the *De purgantibus* by Johannes Damascenus in more than fifteen, and the *De pestilentia* by Rāzī in almost a dozen.

56 Besides the texts known through only one manuscript, see also Anon., *Collectio antidotorum*, ff. 188 verso–208 verso; Bourchan, *Capitulum de capitis dolore*, ff. 106 recto–110 verso; or the several *Lexica plantarum* known through a unique manuscript listed in Touwaide, *Lexica medico-botanica byzantina*.

57 According to Lamagna's publications (supra, n. 26), each and every manuscript of Avicenna, *De urinis*, contains a different version – not only two, contrary to what the titles in the manuscripts suggest (see below).

58 One such collection is by Geôrgios Chioniadês (see Touwaide, *Medicinalia Arabo-Byzantina*, p. 83); and the other by Kônstantinos Meliteniôtês (see *ibid.*, p. 86). We should probably add to these texts the notes *De urinis e Persarum scriptis* (supra, n. 53) and Anon., *Remedia persica (et similia)*, ff. 405 recto–450 verso, 450 verso–453 verso (on which see Costomiris, Etudes, pp. 145–79, 170; and Ieraci Bio, La trasmissione, pp. 133–257, 232). Similarly, we should probably also include in this group the texts identified as coming from Syriac books. Besides the *De urinis e libro syriaco* (supra, n. 53), there are: Anon., *De phlebotomia ex syriaco libro* (MS Bologna, f. 469 verso; MS Città del Vaticano, f. 216 recto); and Anon., *Lexicon botanicum syriacum*, f. 71 verso.

59 Here, and elsewhere, I say "former" because Baghdad had been conquered by the Mongols in 1258.

60 See the cases of Geôrgios Chrysokokkês and Geôrgios Chioniadês in Kunitzsch, Das Fixsternverzeichnis, pp. 382–411; and Pingree, Gregory Chioniades, pp. 133–60.

As for the second group, it consists of translations by individuals identified by name, who translated such authored treatises as Ibn Sīnā's *On urine.*[61] According to their self-identification, these translators were Greeks; and, necessarily, they either knew Arabic or could at least rely on a native speaker collaborator. However, the subsequent history of their translations shows that they may have been natives from areas at the frontier of the two worlds – Byzantium and the former Arabic Empire – who certainly knew both languages, but did not necessarily master them as one would have wished. This is particularly clear in the case of the translation of Ibn Sīnā's *On urine*; indeed, the titles in the manuscripts announce two different versions of the text.[62] The title of one version identifies it as the *second*, claiming that it was necessary because the *other* (i.e., the first) had been made badly, that is, was not in proper Greek.[63] If this story is to be believed, the first translator was bilingual, but without proficiency in either of the two languages! We probably should not over-inflate the historical value of this story by the self-proclaimed *second* and *better* translator. Nevertheless, the interesting point here is that the translators of both versions are individuals identified by name who do not state any connection with an institution – apart from Ioannês Zacharias's title of Aktouarios.[64] Furthermore, if the story presented in the self-proclaimed second translation is to be accepted, it indicates that the first translation was not made in a scholarly milieu but in a bilingual context, most probably by professionals – almost certainly physicians – who aimed to make Arabic treatises of scientific interest available for practical applications by Greek-speaking physicians.[65] Only in a second phase was this first version, which we could qualify as a *first-aid*, better assimilated into the Greek language; and only then did it reach the upper class of Byzantine society,

61 See the works by Lamagna cited above.

62 For these titles, besides Touwaide, *Medicinalia Arabo-Byzantina*, 74–76, see also: *idem*, Arabic Urology, with Lamagna's critique in Per l'edizione del *De urinis*. As mentioned, there are many more versions according to Mario Lamagna, since almost each manuscript presents, unsurprisingly in his opinion, a different version, due to the continual process of revision.

63 The title of this second version states the following (translation mine): "... excellent treatise on urine translated into Greek in a barbaric way by Christodoulos, well-versed in medicine, and now put according to the Greek rhythm and order by the very wise Aktouarios Ioannês Zacharias, well-versed in medicine...."

64 On this title, see Kazhdan, Aktouarios, p. 50.

65 The first translator is identified by the second as Christodoulos, well-versed in medicine (*iatrikôtatos*). We should not overstate the value of this title which may have been used to compensate for the criticism made to Christodoulos, since his work is described as barbaric (*barbarôs*).

thanks to its revision by a learned physician.[66]If this is correct, this second group of thirteenth- and fourteenth-century translations is, with regard to translators, similar to the Siculo-Italian group.

The third group of translations made during this period is very different. It contains heterogeneous material, from lexica of plant names[67] to single notes,[68] collections of formulae for medicines,[69] and short treatises on specific medical topics.[70] For some of them, we probably have the original text of the translation. I shall focus here on these – particularly on bilingual lexica of plant names – as they provide us with many keys for understanding translation agents and agencies.

In one such lexicon, the column on the left contains the Arabic name of the plant transliterated into Greek, and the column on the right its Greek equivalent.[71] The interesting fact is that the two columns were not written by the same hand. Whereas the one on the right is manifestly that of someone used to writing Greek – whom I would call a *native writer* – the one on the left is a hand that hesitates, does not produce regular forms, does not align letters well, and does not use abbreviations in the way of native writers; I would thus suggest identifying this hand as that of a *non-native writer.* On this basis, it seems we might consider such a lexicon the result of a collaboration between a non-native Greek writer, who, instead, is a native Arabic speaker (as he knows the Arabic names of the plants),[72] and a native Greek writer who writes the Greek equivalents of the Arabic names. The codicological analysis of the manuscript

66 The second translator, Iôannês Zacharias, was not only an Aktouarios, but also the author of a medical treatise (Ideler, ed., *De diagnosi*, pp. 353–463). On him, see Scarborough and Talbot, John Aktouarios, pp. 1056.

67 Again, for a list of such lexica see Touwaide, *Lexica medico-botanica byzantina*, pp. 211–28.

68 See (in manuscript) Syrus, *Medicamentum hepaticum*, f. 353 verso; and Avicenna, *De somno*, ff. 177 recto–178 recto.

69 See for example the *Medicamenta* in Eufêmios Sikelios and Filippos Xêros, *Medicamentorum*, ff. 454 recto–464 verso (for the ascription to Filippos Xêros and Eufêmios Sikelios, see Costomiris, Etudes, pp. 170–71; and Ieraci Bio, La trasmissione, pp. 226–27 and 232); Anon., *Anonymi collectio alphabetica remediorum*, ff. 60 recto–73 verso; Anon., *Anonymi collectio remediorum*, f. 450 verso-453 verso; or the *Antidotarium persicum* by Geôrgios Chioniadês and the other by Kônstantinos Meliteniôtês (supra, n. 59).

70 Besides the treatises *De urinis* (supra, n. 53), *De phlebotomia* (supra, n. 59), and the *Capitulum de capitis dolore* (supra, n. 57), see (by the author of the *Capitulum*) Bourchan, *Liber praeservationis sanitatis*, ff. 82 recto–105 recto.

71 Anon., *Lexicon botanicum vocum Arabum*, ff. 202 recto–214 recto.

72 I wish to recall that I am using the term *Arabic* as a convention covering different linguistic and ethnic groups, including e.g., Persians.

containing this lexicon suggests that the codex is from Constantinople and dates back to the first half of the fourteenth century.[73] If our analysis of the hands is correct, this means that Arabic-speaking specialists of medicinal plants were present in Constantinople and interacted with their Greek-speaking colleagues during this period. In other words: we see here what is now called mobility of scientists.

I can now return to the question which, at the conclusion of my analysis of the first group of thirteenth- and fourteenth-century translations, I left unanswered; mainly, the circumstances that made possible what I have called *expeditions* by Byzantine scientists in the territory of the Arabic Empire. The politico-military history of the thirteenth century allows us to suggest a plausible historical reconstruction. Baghdad was conquered in 1258 by the Mongols, who put an end to the ʿAbbāsid Empire;[74] and, shortly after in 1261, the Byzantines recovered their capital, Constantinople, which had been captured by Western troops of the Fourth Crusade in 1204.[75] The sequence of these events in such a short period of time of only three years, and the presence of Byzantines in the territory of the former Arabic Empire as well as of Arabic speaking individuals in Constantinople, suggests bilateral movements of populations: not only Byzantines accessing centers of learning in the former Arabic Empire, but also – if not predominantly – Arabic-speaking scientists emigrating to Byzantium. Our case is more specific: in Constantinople, Arabic-speaking individuals were present in a medical context, together with local physicians, and it is very tempting to hypothesize that these individuals were physicians who left Baghdad and the former ʿAbbāsid territory after the Mongol conquest, moved to Constantinople once it had been recovered from the Latins, and collaborated in the necessary reconstruction of society. This included the recreation of a medical service. At that time, Arabic medicine was the most advanced; moving from Baghdad and the area of the former Arabic Empire to Constantinople in such a time of reconstruction almost certainly provided job opportunities for Arabic-speaking physicians, thanks to their professional competence and skills.[76] According to this view, translations and transfer of knowledge were not necessarily scholarly enterprises – even though they may have included such a component – but had

73 See Touwaide, *Les deux traités*, vol. 1, pp. 97 – 99.

74 See (most recently, and with many references to the literature), Mikaberidze, Baghdad, pp. 174 – 6.

75 See, for example: Treatgold, *History*, pp. 735 – 59, on the period 1261 – 1328.

76 George Saliba (*Islamic Science*, pp. 1 – 129) has made a similar argument about the translators of Greek scientific literature into Arabic during the ninth century, in Baghdad.

more to do with practical necessities and what we nowadays would call humanitarian assistance.

Also according to such a view, thirteenth- and fourteenth-century translations were probably not centralized in a definite place, but were most plausibly made wherever medical services were functioning. No available evidence allows us to confirm or deny this assumption. However, by the mid-fourteenth century, the interest in Arabo-Islamic medicine and, hence, translation from Arabic (or Persian) into Greek, seems indeed to have been concentrated in one place: the monastery of Ioannes Prodromenos (St. John the Precursor) in Constantinople,[77] the scientific activity of which was particularly illustrated at that time by Neophytos Prodromenos.[78]

A polymath and still-enigmatic figure like Symeon Seth, Neophytos was a philosopher, a naturalist, and also a physician. Supposedly of Albanian origin, he appears to have been interested in Arabo-Islamic medicine, including the *Efodia*.[79] Characteristically, the interest which seems to have crystalized around him goes together with a revival of the Greek medical tradition, particularly Dioscorides and the *De materia medica*.[80] At first glance, this duality would invite us to consider that there was a clash of cultures and that the reintroduction of Greek medical literature was a response to the presence of its Arabic equivalent. A closer examination of extant material does not confirm such an immediate interpretation. On the contrary, there seems to be a continuity between Arabic and Greek literature, mainly thanks to bilingual lexica. Such works made it possible, indeed, to connect the two textual bodies and to highlight their complementarity. Does this mean that the monastery of the Prodromos was a translation agency? I am not sure of this, and I think rather that it was a center for reorganization of previously translated works, so as to create a unified body out of pieces of different origins. I am all the more inclined to propose such an interpretation because the monastery is known to have been made of a scriptorium, a library, a school –

---

77 On the Prodromos monastery, see Talbot, Petra Monastery, p. 1643; and Allen, Stefan Uroš II Milutin, pp. 1949–50.

78 On Neophytos, see Hunger, *Hochsprachliche profane Literatur*, pp. 308–309; Eftychiadou, *Eisagôgê*, p. 211; and Cacouros, Néophytos Prodromênos, pp. 193–212. On the logical works of Neophytos, see Cacouros, Deux épisodes inconnus, pp. 589–627. On his medical work, see Touwaide, Un recueil de pharmacologie, pp. 13–56.

79 He compiled a lexicon of "Saracen" terms translated from the *Efodia* which can be read in a manuscript at the Bibliothèque nationale de France (*graecus* 2286, f. 54 recto). This is a very short lexicon that seems to be either a fragment or the beginning of a work that was not pursued further.

80 On this point, see, for instance, Touwaide, Development of Paleologan Renaissance, pp. 189–224.

which later became a university, and even the last of the Byzantine empire – and a hospital.[81] This being the case, I suspect that the production of books in the scriptorium and texts in the school may have been linked with the activity in the hospital. In other words: the complex of the Prodromos was not only a learning center, but also – if not primarily – a center that brought together the most advanced science of that time (Arabo-Islamic medicine), and the local tradition (Greek medicine), to merge and apply them in a productive way in its very own hospital. If so, we again encounter the notion of a practical application of the translations of Arabic (or Persian) medical treatises into Greek, which we have suspected throughout our analyses.

5 Conclusions

What do Byzantine versions of Arabic medical literature tell us about translation agents and agencies? Both a great deal and not as much as we would like. More than anything else, they appear to show us individuals of different ethnic and linguistic backgrounds associated in the exercise of their profession and working in close collaboration for practical purposes. They also suggest that translation was not necessarily an intellectual exercise *per se*, but may have been a necessity in the management of individual lives and societies, so as to make it possible to take advantage of the most recent and advanced science.

If such an assessment is correct, the translations under consideration here invite us to consider the agents more than the agencies, and suggest that we should be cautious about theoretical models which create schools where there are only – in a certain sense – individuals at work, however vast and important the programs they were engaged in may have been. At the same time, the translations I have here been dealing with also reveal the limitations of our documentation and, hence, the fragility of our analyses, even though the manuscripts containing such translations provide a great deal of useful information. We have no other recourse than formulating hypotheses. However plausible they

81 On the scriptorium in the fourteenth century, see Touwaide, Un recueil de pharmacologie. On the library, see Volk, *Die byzantinischen Klosterbibliotheken*; Kakoulidê, Ê bibliothêkê, pp. 3–39; and Touwaide, op. cit. On the school of Prodromos, see Fuchs, *Die höheren Schulen*; and, for the fifteenth century: Mondrain, Jean Argyropoulos, pp. 233–50; and as for its being the last university of the Byzantine Empire, see, for example, Kazhdan, University of Constantinople, p. 2143. Finally, on the *Xenodocheion tou Kralê* at the Prodromos, see Talbot, Xenon of the Kral, p. 2209.

may be, our reconstructions are tentative and will remain so until new and currently unexpected documentation comes to light.

Bibliography

ʿAbd Allāh, *Tractatus medicae artis*, MS Vienna, Österreichische Nationalbibliothek, *medicus graecus* 21, ff. 1 recto–81 recto.

J.S. Allen, Stefan Uroš II Milutin, in Alexander P. Kazhdan, Alice Mary Talbot, Anthony Cutler, Timothy E. Gregory, and Nancy P. Ševčenko, eds., *The Oxford Dictionary of Byzantium*, 3 vols., New York and Oxford: Oxford University Press, 1991, pp. 1949–50.

Filius Ameroumanae Afri (*uios tou amerourmnê afrikês*), MS Oxford, Bodleian Library, *Codex Baroccianus* 150, ff. 17 verso–18 recto.

S. Ammar, *Ibn Al Jazzar & the Medical School of Kairouan*,s.l. n.d.

Anon., *Collectio alphabetica remediorum, initio mutila*, MS Paris, Bibliothèque nationale de France, *graecus* 2219, ff. 60 recto–73 verso.

Anon., *Collectio remediorum ex libris Persicis*, MS Paris, Bibliothèque nationale de France, *graecus* 2194, ff. 450 verso–453 verso.

Anon., *Collectio antidotorum*, MS Bologna, Biblioteca Universitaria 3632, ff. 188 verso–208 verso.

Anon., *Lexicon botanicum vocum Arabum*, MS Paris, Bibliothèque nationale de France, *graecus* 2287, ff. 202 recto–214 recto.

Anon., *Lexicon botanicum syriacum*, MS London, Wellcome Library, MSL 60, f. 71 verso.

Anon., *De phlebotomia ex syriaco libro*, MS Bologna, Biblioteca Universitaria 3632, f. 469 verso; and MS Vatican City, Biblioteca Apostolica Vaticana, *Palatinus graecus* 279, f. 216 recto.

Anon., *Remedia persica* (*et similia*), MS Paris, Bibliothèque nationale de France, *graecus* 2194, ff. 405 recto–450 verso, 450 verso–453 verso.

Anon., *De urinis e Persarum scriptis*, MS Paris, Bibliothèque nationale de France, *graecus* 2309, ff. 41 recto–43 verso.

Ataamatoulmardaou, *Dieta infirmorum*, MS Vienna, Österreichische Nationalbibliothek, *medicus graecus* 21, ff. 186 recto–206 verso.

Avicenna, *De somno*, MS Paris, Bibliothèque nationale de France, *graecus* 2260, ff. 177 recto–178 recto.

B. Baldwin, Synesius of Cyrene, in Alexander P. Kazhdan, Alice Mary Talbot, Anthony Cutler, Timothy E. Gregory, and Nancy P. Ševčenko, eds., *The Oxford Dictionary of Byzantium*, 3 vols., New York and Oxford: Oxford University Press, 1991, p. 1993.

M.-G. Balty-Guesdon, Le "Bayt al-Hikma" de Baghdad, *Arabica* 29, 1992, pp. 131–50.

M.-G. Balty-Guesdon, *Bayt al-Hikma* et politique culturelle du Calife Al-Ma'mûn, in Luciana Rita Angeletti and Alain Touwaide, eds., *Medieval Arabic Medicine*, 2 vols. (= *Medicina nei Secoli*, 6,2, 1994, & 7,1, 1995), Rome: Delfino, vol. 1, pp. 275–91.

J.S. Bernard, *Synesius de febribus, Quem nunc primum ex codice MS. Bibliothecae Lugduno Batavae edidit, vertit, notisque illustravit. Accedit Viaticum Constantino Africano interprete lib. VII.* Pars. Amstelaedami: Apud Gerard de Groot, et Lugduni Batavorum: Apud Philippum Bonk, 1749.

G. Bos (transl.) and Ibn al-Ğazzār, *Ibn al-Jazzār on sexual diseases and their treatment: a critical edition of Zād al-musāfir wa-qūt al-ḥāḍir: Provisions for the traveller and nourishment for the sedentary: book 6: the original Arabic text with an English translation, introduction and commentary by Gerrit Bos,* London; New York: Kegan Paul International, 1997.

P. Bouras-Vallianatos, Galen's Reception in Byzantium: Symeon Seth and his Refutation of Galenic Theories on Human Physiology, *Greek, Roman, and Byzantine Studies* 55, 2015, pp. 431–469.

Bourchan, *Capitulum de capitis dolore*, MS Vienna, Österreichische Nationalbibliothek, *medicus graecus* 21, ff. 106 recto–110 verso.

Bourchan, *Liber praeservationis sanitatis*, MS Vienna, Österreischische Nationalbibliothek, *medicus graecus* 21, ff. 82 recto–105 recto.

M. Brunet, *Symeon Seth, médecin de l'empereur Michel Doucas. Sa vie, son oeuvre,* Bordeaux: Delmas, 1939.

M. Cacouros, Deux épisodes inconnus dans la réception de Proclus à Byzance aux XIIIe–XIVe siècles: la philosophie de Proclus réintroduite à Byzance grâce à l'hypotypôsis. Néophytos Prodromènos et Kôntostéphanos (?) lecteurs de Proclus (avant Argyropoulos) dans le Xénôn du Kralj, in Alain Ph. Segonds and Carlos Steel, eds., *Proclus et la théologie platonicienne, Actes du Colloque International de Louvain (13–16 mai 1998 en l'honneur de H. D. Saffrey et L. G. Westerink)*, Leuven: University Press; and Paris: Belles Lettres, 2000, pp. 589–627.

M. Cacouros, Néophytos Prodromènos copiste et responsable (?) de l'édition quadrivium-corpus aristotelicum du 14e siècle, *Revue des études byzantines* 56, 1998, pp. 193–212.

P. Canart, Le livre grec en Italie méridionale sous les règnes normands et souabes, *Scrittura e civiltà* 2, 1978, pp. 103–162.

G. Cavallo, La trasmissione scritta della cultura greca antica in Calabria e in Sicilia tra i secoli X–XV. Consistenza, tipologia, fruizione, *Scrittura e civiltà* 4, 1980, pp. 157–245.

G. Cavallo, La cultura italo-greca nella produzione libraria, in Guglielmo Cavallo, Vera von Falkenhausen, Raffaella Farioli Campanati, Marcello Gigante, Valentino Pace, and Franco Panvini Rosati, eds., *I Bizantini in Italia*, Milan: Libri Schweiwiller, 1982, pp. 495–612.

J. Channing, *Rhazes de Variolis et Morbilis, Arabic et Latin; cum Aliis Nunnulis Eiusdem Argumenti.* Cura et impensis – ..., Londini: Excudebat Guilielmus Bowyer, 1766.

H. Condylis-Bassoukos, *Kalīla wa-Dimna:* le chapitre du jugement de *Dimna* et sa traduction en grec. De l'acculturation du politique, *Byzantinoslavica* 65, 2007, pp. 41–50.

H. Condylis-Bassoukos, À propos de la traduction grecque de *Kalila wa-Dimna, Graeco-Arabica* 3, 1984. pp. 207–12.

H. Condylis-Bassoukos, Stéphanitès kai Ichnélatès, traduction grecque de *Kalila wa-Dimna, Le Muséon* 103, 1990, pp. 139–49.

H. Condylis-Bassoukos, *Stéphanitès kai Ichnélatès, traduction grecque (XIe siècle) du livre Kalīla wa-Dimna d'Ibn al-Muqaffa' (VIIIe siècle), Étude lexicologique et littéraire* (Fonds Draguet de l'Académie Royale des Sciences, des Lettres et des Beaux-Arts de Belgique, Classe des Lettres, 11), Leuven and Paris: Peeters, 1997.

M.-H. Congourdeau, Le monde byzantin, in *À l'ombre d'Avicenne. La médecine au temps des califes. Exposition présentée du 18 novembre 1996 au 2 mars 1997,* Paris: Institut du Monde Arabe; Gand: Snoeck-Ducaju & Zoon, 1996, pp. 271–73.

M.-H. Congourdeau, A propos d'un chapitre des *Ephodia:* l'avortement chez les médecins grecs, *Revue des etudes byzantines* 55, 1997, pp. 261–77.

M.-H. Congourdeau, Le traducteur grec du traité de Rhazès sur la variole, in Antonio Garzya and Jacques Jouanna, eds., *Storia e ecdotica dei testi medici greci. Atti del II Convegno internazionale, Parigi, 24–26 maggio 1994* (Collectanea 10), Napoli: D'Auria, 1996, pp. 99–111.

G.A. Costomiris, Etudes sur les écrits inédits des anciens médecins grecs. Deuxième série: L'Anonyme de Daremberg, Métrodora, Aétius, *Revues des études grecques* 3, 1890, pp. 145–79, 170.

H.O. Coxe, *Catalogi Codicum Manuscriptorum Bibliothecae Bodleianae.* Pas Prima Recensionem Codicum Graecorum Continens, Oxonii: E Typographeo Academico, 1853.

Ch. Daremberg, Notices et extraits des manuscrits médicaux grecs et latins des principales bibliothèques d'Angleterre, *Archives et Missions scientifiques et littéraires*, Ière série, 2, 1851, pp. 113–68, 470–1, 484–548; and 3, 1854, pp. 1–51. Reprinted as *Notices et extraits des manuscrits médicaux grecs, latins et français des principales bibliothèques d'Europe*, Ière partie: Manuscrits grecs d'Angleterre, Paris: Imprimerie impériale, 1854.

A. Delatte, ed., *Anecdota atheniensia et alia*, vol. 2, Paris: Belles Lettres, 1939.

H. Diels, ed., *Die Handschriften der antiken Ärzte. II. Teil. Die übrigen griechische Ärzte ausser Hippokrates und Galenos (= Abhandlungen der Königlichen Preussischen Akademie der Wissenschaften, Philosophisch-historische Klasse, Jahre 1906, Abhandlung I*), Berlin: Königliche Akademie der Wissenschaften, 1906. (Published under a separate title as: *Idem*, ed., *Die Handschriften der antiken Ärzte. Griechische Abteilung*, Berlin: Königliche Akademie der Wissenschaften, 1906). Supplement in *idem*, *Bericht über den Stand des interakademischen Corpus medicorum antiquorum und erster Nachtrag zu den in den Abhandlungen 1905 und 1906 veröffentlichten Katalogen, Die Handschriften der antiken ärzte I. und II. Teil (= Abhandlungen der Königlichen Preussischen Akademie der Wissenschaften, Philosophisch-historische Klasse, Jahre 1907, Abhandlung II*), Berlin: Königliche Akademie der Wissenschaften, 1908.

T.-A. Druart, Razi, in Thomas Glick, Steven J. Livesey, and Faith Wallis, eds., *Medieval Science, Technology, and Medicine: An Encyclopedia*, New York and London: Routledge, 2005, pp. 434–36.

J. Duffy (ed. and transl.) and Johannes Alexandrinus, *Ioannis Alexandrini In Hippocratis Epidemiarum librum VI commentarii fragmenta* [...] (Corpus Medicorum Graecorum XI 1,4), Berlin: Akademie Verlag, 1997.

A. Eftychiadou, *Eisagôgê eis tên byzantinên therapeutikên*, Athens: Parisianos, 1983.

E. Fisher, Greek Translations of Latin Literature in the Fourth Century AD, *Yale Classical Studies* 27, 1982, pp. 173–216.

C. Förstel, Alexandre de Tralles, Thérapeutique, in *Byzance retrouvée. Erudits et voyageurs français (XVIe–XVIIIe siècles). Chapelle de la Sorbonne-Paris, 13 août–2 septembre 2001*, Paris: Centre d'études byzantines, néo-helléniques et sudesteuropéennes-EHESS; Publications de la Sorbonne, Byzantina Sorboniensia, 2001, pp. 27–30.

M.B. Foti, *Il Monastero del S.mo Salvatore in lingua phari. Proposte scrittorie e coscienza culturale*, Messina: n.p., 1989.

F. Fuchs, *Die höheren Schulen von Konstantinopel im Mittelalter* (Byzantinisches Archiv 8), Leipzig and Berlin: B.G. Teubner, 1926.

F. Gabrieli, Il 'Zâd al Musâfir' di Ibn al Gazzâr in un ms. Greco Corsiniano (ΕΦΟΔΙΑ ΤΟΥ ΑΠΟΔΗΜΥΝΤΟΣ), *Rendiconti della Reale Accademia dei Lincei*, Classi di scienze morali, storiche e filologiche, serie V, 14, 1905, pp. 205–20.

J. Goupyl, *Alexandri Tralliani Medici. Libri XII Rhazae De pestilentia libellus ex Syrorum lingua in Graecam translatus. Jacopi Goupyli in eosdem castigationes ...*, Lutetiae: Ex officina Rob Stephani, 1548.

M.H. Green, Constantine the African, in Thomas Glick, Steven J. Livesey, and Faith Wallis, eds., *Medieval Science, Technology, and Medicine: An Encyclopedia*, New York and London: Routledge, 2005, pp. 145–47.

W.A. Greenhill, *A Treatise on the Small-pox and Measles, by Abú Becr Mohammed ibn Zacaríyá ar-Rází (commonly called Rhazes)*, Translated from the original Arabic, London: Printed for the Sydenham Society, 1848.

D. Gutas, *Greek Thought, Arabic Culture: The Graeco-Roman Translation Movement in Baghdad and Early 'Abbāsid Society (2nd–4th/8th–10th centuries)*, London and New York: Routledge, 1998.

J. Hamesse, ed., *Les traducteurs au travail: Leurs manuscripts et leurs methods*, Actes du Colloque international organisé par le Ettore Majorana Centre for Scientific Culture (Erice 30 septembre–6 octobre 1999), Turnhout: Brepols, 2001.

G. Harig, Von den arabischen Quellen des Simeon Seth, *Medizinhistorisches Journal* 2, 1967, pp. 248–268.

H. Hunger, *Die hochsprachliche profane Literatur der Byzantiner*, Zweiter Band: Philologie, Profandichtung, Musik, Mathematik und Astronomie, Naturwissenschaften, Medizin, Kriegswissenschaften, Rechtsliteratur (= *Handbuch der Altertumswissenschaft*, Zwölfte Abteilung: *Byzantinisches Handbuch*, Fünfter Teil, Zweiter Band), Munich: C. H. Beck'sche Verlagsbuchhandlung, 1978.

S.P. Huntington, *The Clash of Civilizations and the Remaking of World Order*, New York: Simon and Schuster, 1996.

J.L. Ideler, ed., *De diagnosi*, in *idem*, ed., *Physici et medici Graeci minores. Congessit, ad fidem codd. Mss. praesertim eorum, quos beatus Dietzeius contulerat, veterumque editionum partim emendavit partim nunc prima vice edidit, commentariis criticis indicibusque tan rerum quam verborum instruxit*, 2 vols., Berlin: Typis et impensis G. Reimeri, 1841–1842, vol. 2, pp. 353–463.

J.L. Ideler, ed., *De urinis e libro syriaco*, in *idem*, ed., *Physici et medici Graeci minores. Congessit, ad fidem codd. Mss. praesertim eorum, quos beatus Dietzeius contulerat, veterumque editionum partim emendavit partim nunc prima vice edidit, commentariis criticis indicibusque tan rerum quam verborum instruxit*, 2 vols., Berlin: Typis et impensis G. Reimeri, 1841–1842, vol. 2, pp. 303–4.

J.L. Ideler, ed., *De urinis ex Persarum disciplina*, in *idem*, ed., *Physici et medici Graeci minores. Congessit, ad fidem codd. Mss. praesertim eorum, quos beatus Dietzeius contulerat, veterumque editionum partim emendavit partim nunc prima vice edidit, commentariis criticis indicibusque tan rerum quam verborum instruxit*, 2 vols., Berlin: Typis et impensis G. Reimeri, 1841–1842, vol. 2, pp. 305–306.

A.M. Ieraci Bio, La trasmissione della letteratura medica greca nell'Italia meridionale fra X e XV secolo, in Antonio Garzya, ed., *Contributi alla cultura greca nell'Italia meridionale* I (Hellenica et Byzantina Neapolitana 13), Naples: Bibliopolis, 1989, pp. 133–257, 232.

Isaac Israelitus, *De elaterio*, MS Vatican City, Biblioteca Apostolica Vaticana, *graecus* 300, f. 296 recto.
A.Z. Izkandar, Razi, in Helaine Selin, ed., *Encyclopaedia of the History of Science, Technology, and Medicine in Non-Western Cultures*, 2 vols., Berlin, Heidelberg, New York: Springer, 2008, vol. 1, pp. 155–56.
H.D. Kakoulidê, Ê bibliothêkê tou Monês Prodromou Petras stên Kônstantinoupolê, *Ellênika* 21, 1968, pp. 3–39.
A.P. Kazhdan, Aktouarios, in Alexander P. Kazhdan, Alice Mary Talbot, Anthony Cutler, Timothy E. Gregory, and Nancy P. Ševčenko, eds., *The Oxford Dictionary of Byzantium*, 3 vols., New York and Oxford: Oxford University Press, 1991, p. 50.
A.P. Kazhdan, Romanos II, in Alexander P. Kazhdan, Alice Mary Talbot, Anthony Cutler, Timothy E. Gregory, and Nancy P. Ševčenko, eds., *The Oxford Dictionary of Byzantium*, 3 vols., New York and Oxford: Oxford University Press, 1991, pp. 1806–1807.
A.P. Kazhdan, Seth, Symeon, in Alexander P. Kazhdan, Alice Mary Talbot, Anthony Cutler, Timothy E. Gregory, and Nancy P. Ševčenko, eds., *The Oxford Dictionary of Byzantium*, 3 vols., New York and Oxford: Oxford University Press, 1991, pp. 1882–83.
A.P. Kazhdan, University of Constantinople, in Alexander P. Kazhdan, Alice Mary Talbot, Anthony Cutler, Timothy E. Gregory, and Nancy P. Ševčenko, eds., *The Oxford Dictionary of Byzantium*, 3 vols., New York and Oxford: Oxford University Press, 1991, p. 2143.
A. Kouzis, Quelques considérations sur les traductions en grec des oeuvres médicales orientales et principalement sur les deux manuscrits de la traduction d'un traité persan par Constantin Méliténiotis, *ΠΡΑΚΤΙΚΑ ΤΗΣ ΑΚΑΔΗΜΙΑΣ ΑΘΗΝΩΝ* 14, 1939, pp. 205–20.
P. Kunitzsch, Das Fixsternverzeichnis in der "Persischen Syntaxis" des Georgios Chrysokokkes, *Byzantinische Zeitschrift* 52, 1964, pp. 382–411.
M. Lamagna, Per l'edizione del *De urinis* attribuito ad Avicenna : studio complessivo della tradizione manoscritta, *Revue d'histoire des texts* n.s. 6, 2011, pp. 15–26.
M. Lamagna, La *recensio amplior* del *De urinis* di Avicenna: lo stato della tradizione manoscritta, in Véronique Boudon-Millot, Antonio Garzya, Jacques Jouanna, Amneris Roselli, eds., *Ecdotica e Ricezione dei testi medici greci. Atti del V Convegno Internazionale, Napoli, 1–2 ottobre 2004*, Naples: M. D'Auria Editore, 2006, pp. 321–33.
M. Lamagna, La *recensio amplior* inedita del *De urinis* di Avicenna, in Antonio Garzya and Jacques Jouanna, eds., *Trasmissione e Ecdotica dei testi medici greci. Atti del VI Convegno Internazionale, Parigi, 17–19 maggio 2001*, Naples: M. D'Auria Editore, 2003, pp. 271–80.
B. Langkavel, ed., *Simeoni Sethii, De alimentorum facultatibus*, Leipzig: B.G. Teubner, 1868.
K.-H. Leven, Seth, Symeon, in Wolfgang-Uwe Eckart and Christoph Gradmann, eds., *Ärztelexikon, Von der Antike bis zum 20. Jahrhundert,* Beck'sche Reihe, Munich: Beck, 1995, pp. 330–31.
K.-H. Leven, Zur Kenntnis der Pocken in der arabischen Medizin, im lateinischen Mittelalter und in Byzanz, in Odilo Engels und Peter Schreiner, eds., *Die Begegnung des Westens mit dem Osten.* Kongressakten des 4. Symposions des Mediävistenverbandes in Köln aus Anlass des 1000. Todesjahres der Kaiserin Theophanu, Sigmaringen: Jan Thorbecke Verlag, 1993, pp. 341–54.
B. Lewis, The Roots of Muslim Rage, in *The Atlantic Monthly*, September 1990, pp. 47–60.
S. Lucà, I Normanni e la 'Rinascita' del sec. XII, *Archivio storico per la Calabria e la Lucania* 60, 1993, pp. 36–63.

M. Mavroudi, Exchanges with Arabic Writers during the Late Byzantine Period, in Sarah T. Brooks, ed., *Byzantium: Faith and Power (1261–1557): Perspectives on Late Byzantine Art and Culture*, New York: The Metropolitan Museum of Art; New Haven & London: Yale University Press, 2006, pp. 62–75.

G. Mercati, Filippo Xeros reggino, Giovanni Alessandrino iatrosofista e altri nel codice vaticano degli *Ephodia*, in *idem*, *Notizie varie di antica letteratura medica* (Studi e testi 31), Rome: Tipographia Poliglotta Vaticana, 1917, pp. 9–17.

I. Mercati and P. Franchi de' Cavalieri, *Bybliothecae Apostolicae Vaticanae codices manuscripti recensiti ... Codices Vaticani graeci*, Tomus I. Codices 1–329, Rome: Typis Polyglottis Vaticanis, 1926.

F. Micheau, La connaissance d'ibn al-Jazzâr, médecin de Kairouan, dans l'Orient arabe, *Arabica* 43, 1996, pp. 385–405.

A. Mikaberidze, Baghdad, Siege of – (1258), in Alexander Mikaberidze, ed., *Conflict and Conquest in the Islamic World*, 2 vols., Santa Barbara: ABC-Clio, 2011, vol. 1, pp. 174–6.

J. Mogenet, L'influence de l'astronomie arabe à Byzance du IXe *au XIV*e siècle, in *Colloques d'Histoire des Sciences, I (1972) & II (1973) organisés par le Centre d'Histoire des Sciences et des Techniques de l'Université Catholique de Louvain* (Université de Louvain, Recueil de travaux d'histoire et de philologie, 6e Série, Fascicule 9), Louvain: Université catholique de Louvain, 1976, pp. 44–55.

J. Mogenet, Une scolie inédite du Vat. gr. 1594 sur les rapports entre l'astronomie 16 arabe et Byzance, *Osiris* 14, 1952, pp. 198–221.

B. Mondrain, Jean Argyropoulos professeur à Constantinople et ses auditeurs médecins, d'Andronic Eparque à Démétrios Angelos, in Cornula Scholz and Georgios Makris, eds., *Polypleuros nous. Miscellanea für Peter Schreiner zu Seinem 60. Geburtstag* (Byzantinisches Archiv 19), Munich: K.G. Saur, 2000, pp. 233–50.

V. Nutton, Jundîshâbûr, in *A l'ombre d'Avicenne. La médecine au temps des Califes. Exposition présentée du 18 novembre 1996 au 2 mars 1997 [Paris, Institut du Monde Arabe]*, Paris: Institut du Monde Arabe, and Gand: Snoeck & Ducajou, 1996, p. 22.

J.-M. Olivier, *Répertoire des bibliothèques et catalogues de manuscrits grecs de Marcel Richard*, Troisième édition entièrement refondue (Corpus Chistianorum, Series graeca), Turnhout: Brepols, 1995.

D. Pingree, Gregory Chioniades and Palaeologan Astronomy, *Dumbarton Oaks Papers* 18, 1964, pp. 133–60.

P.E. Pormann and E. Savage-Smith, *Medieval Islamic Medicine*, The New Edinburgh Islamic Surveys, Edinburgh: Edinburgh University Press, 2007.

Neophytos Prodromenos, [lexicon of "Saracen" terms translated from the *Efodia*], MS Paris, Bibliothèque nationale de France, *graecus* 2286, f. 54 recto.

M. Re, A proposito dello 'Skylitzes' di Madrid, *La memoria. Annali della Facoltà di Lettere e Filosofia dell'Università di Palermo* 3 , 1984, pp. 329–341.

L. Richter-Bernburg, Razi, in Josef W. Meri, ed., *Medieval Islamic Civilization: An Encyclopedia*, 2 vols., New York and London: Routledge, 2006, vol. 2, pp. 671–74.

G. Saliba, *Islamic Science and the Making of the European Renaissance*, Boston: MIT Press, 2007.

J. Scarborough and A.M. Talbot, John Aktouarios, in Alexander P. Kazhdan, Alice Mary Talbot, Anthony Cutler, Timothy E. Gregory, and Nancy P. Ševčenko, eds., *The Oxford Dictionary of Byzantium*, 3 vols., New York and Oxford: Oxford University Press, 1991, pp. 1056.

M. Schmid, Eine *Galen-Kontroverse* des Simeon Seth, in XVIIe Congrès International d'Histoire de la Médecine, Athènes – Cos, 4–14 septembre 1960, Communications, Athens, 1960, vol. 1, pp. 491–5.

F. Sezgin, *Geschichte des arabischen Schrifttums*, Band 3: Medizin, Pharmazie, Zoologie, Tierheilkunde bis ca. 430 H., Leiden: E.J. Brill, 1971.

Eufêmios Sikelios and Filippos Xêros, *Medicamentorum compositiones ab Euphemio Siculo et Philippo Xero Rheginensi*, MS Paris, Bibliothèque nationale de France, *graecus* 2194, ff. 454 recto–464 verso.

R.E. Sinkewicz and W.M. Hayes, *Manuscript Listings for the Authored Works of the Palaeologan Period* (Greek Index Project Series, 2), Toronto: Pontifical Institute for Mediaeval Studies, 1989.

R.E. Sinkewicz and W.M. Hayes, *Manuscript Listings for the Authors of Classical and Late Antiquity* (Greek Index Project Series, 3), Toronto: Pontifical Institute for Mediaeval Studies, 1990.

R.E. Sinkewicz and W.M. Hayes, *Manuscript listings for the Authors of the Patristic and Byzantine Periods* (Greek Index Project Series, 4), Toronto: Pontifical Institute for Mediaeval Studies, 1992.

L.-O. Sjöberg, *Stephanites und Ichnelates. Überlieferungsgeschichte und Text* (Studia Graeca Upsaliensia 2), Stockholm: Almqvist & Wiksell, 1962.

Abraham Syrus [of the Mangana hospital], *Medicamentum hepaticum*, MS Florence, Biblioteca Medicea Laurenziana, Antinori 101, f. 353 verso.

A.M. Talbot, Petra Monastery, in Alexander P. Kazhdan, Alice Mary Talbot, Anthony Cutler, Timothy E. Gregory, and Nancy P. Ševčenko, eds., *The Oxford Dictionary of Byzantium*, 3 vols., New York and Oxford: Oxford University Press, 1991, p. 1643.

A.M. Talbot, Xenon of the Kral, in Alexander P. Kazhdan, Alice Mary Talbot, Anthony Cutler, Timothy E. Gregory, and Nancy P. Ševčenko, eds., *The Oxford Dictionary of Byzantium*, 3 vols., New York and Oxford: Oxford University Press, 1991, p. 2209.

A. Tihon, Les tables astronomiques persanes à Constantinople dans la première moitié du XIVe siècle, *Byzantion* 57, 1987, pp. 603–24. Reprinted in *eadem*, *Etudes d'astronomie Byzantine* (Variorum Collected Studies Series CS454), Aldershot: Variorum, 1994.

A. Tihon, Tables islamiques à Byzance, *Byzantion* 60, 1960, pp. 401–25.

A. Tihon, Les textes astronomiques arabes importés à Byzance aux XI et XII siècles, in Baudouin van den Abeele, Anne Tihon, and Isabelle Draelants, eds., *Occident et Proche-Orient. Contacts scientifiques au temps des Croisades. Actes du colloque de Louvain-la-Neuve, 24 et 25 mars 1997 (Réminiscences* 5), Turnhout: Brepols, 2000, pp. 313–24.

A. Touwaide, *Ancient Botany from Byzantium to the West: Catalogue of the Exhibition held at Dumbarton Oaks, Washington, DC, May 1–July 10, 2000*, Washington, DC: Dumbarton Oaks Center for Byzantine Studies, 2000.

A. Touwaide, Arabic into Greek: The Rise of an International Lexicon of Medicine in the Medieval Eastern Mediterranean?, in Robert Wisnovsky, Faith Wallis, Jamie C. Fumo, and Carlos Fraenkel, eds., *Vehicles of Transmission, Translation, and Transformation in Medieval Textual Culture* (Cursor Mundi 4), Turnhout: Brepols, 2011, pp. 195–222.

A. Touwaide, Arabic Materia Medica in Byzantium during the 11th century A.D. and the problems of transfer of knowledge in Medieval science, in S. Mohammad Razaullah Ansari, ed., *Science and Technology in the Islamic World.* Proceedings of the XXth

International Conference of History of Science, Liège, 20 – 26 July 1997, vol. XX (*De Diversis Artibus*. Collection of Studies from the International Academy of the History of Science 64), Turnhout: Brepols, 2002, pp. 223 – 46.

A. Touwaide, Arabic Medicine in Greek Translation: A Preliminary Report, *Journal of the International Society for the History of Islamic Medicine* 1, 2002, pp. 45 – 53.

A. Touwaide, Arabic Urology in Byzantium, in Natale Gaspare De Santo, Luigi Iorio, Spyprs G. Marketos, Shaul G. Massry and Gary Eknoyan, eds., *The History of Nephrology, New Series* 1, Milan: Wichtig, 2004, pp. 167 – 73.

A. Touwaide, Botany and Humanism in the Renaissance: Background, Interaction, Contradictions, in Therese O'Malley and Amy Meyers, eds., *The Art and History of Botanical Painting and Natural History Treatises*, Washington, DC: The National Gallery of Art, 2008, pp. 33 – 61.

A. Touwaide, Byzantine Medical Manuscripts: Towards a New Catalogue, with a Specimen for an Annotated Checklist of Manuscripts Based on an Index of Diels' Catalogue, *Byzantion* 79, 2009, pp. 453 – 595.

A. Touwaide, Les deux traités de toxicologie attribués à Dioscoride. La tradition manuscrite grecque, édition critique du texte grec et index, 5 vols., PhD diss., Université catholique de Louvain, 1981.

A. Touwaide, The Development of Paleologan Renaissance: An analysis based on Dioscorides' *De materia medica*, in Michel Cacouros and Marie-Hélène Congourdeau, eds., *Philosophie et sciences à Byzance de 1204 à 1453, Actes de la Table Ronde organisée au XXe Congrès International d'Études Byzantines, Paris, 2001* (Orientalia Lovaniensia Analecta 146), Leuven, Paris and Dudley: Peeters and Department Oosterse Studies Leuven, 2006, pp. 189 – 224.

A. Touwaide, Enfermedad y curación, in Alain Touwaide, ed., *Herbolarium et material medica. Libro de Estudios (Biblioteca Statale de Lucca, ms. 296). Facsimile y estudio del manuscrito 296 de la Biblioteca Civica de Lucca*, Madrid: Arte y Naturaleza Ediciones; and Lucca: Biblioteca Statale di Lucca, 2007, 2: 155 – 66.

A. Touwaide, Las fuentes de la terapia medieval, in Alain Touwaide, ed., *Herbolarium et material medica. Libro de Estudios (Biblioteca Statale de Lucca, ms. 296). Facsimile y estudio del manuscrito 296 de la Biblioteca Civica de Lucca*, Madrid: Arte y Naturaleza Ediciones; and Lucca: Biblioteca Statale di Lucca, 2007, 2: 29 – 41.

A. Touwaide, Greek Medical Manuscripts: Toward a New Catalogue, *Byzantinische Zeitschrift* 101, 2008, pp. 199 – 208.

A. Touwaide, Healers and Physicians in Ancient and Medieval Cultures, in Zohara Yaniv and Uriel Bachrach, eds., *Handbook of Medicinal Plants*, Binghamton: Haworth Medical Press, 2005, pp. 155 – 73.

A. Touwaide, L'intégration de la pharmacologie grecque dans le monde arabe. Une vue d'ensemble, in Luciana R. Angeletti and Alain Touwaide, eds., *History of Arabic Medicine*, 2 vols. (= *Medicina nei Secoli* 6.2, 1994 and 7.1, 1995), Rome: Delfino, 2: 259 – 89.

A. Touwaide, The Jujube Tree in the Eastern Mediterranean: A Case Study in the Methodology of Textual Archeobotany, in Peter Dendle and Alain Touwaide, eds., *Health and Healing from the Medieval Garden*, Woodbridge: Boydell and Brewer, 2008, pp. 72 – 100.

A. Touwaide, Kidney Dysfunction, from the Arabic to the Byzantine World in 11th- and 12th-Century Southern Italy, in Natale Gaspare De Santo, Shaul G. Massry, and Alain

Touwaide, eds., *History of Nephrology – Proceedings of the 6th Conference of the International Association for the History of Nephrology. Special Issue of Journal of Nephrology*, Milan: Wichtig, 2009, pp. 12–20.

A. Touwaide, The Legacy of Classical Antiquity in Byzantium and the West, in Peter Dendle and Alain Touwaide, eds., *Health and Healing from the Medieval Garden*, Woodbridge: Boydell and Brewer, 2008, pp. 15–28.

A. Touwaide, *Lexica medico-botanica byzantina*. Prolégomènes à une etude, in *Tês filiês tade dôra – Miscelánea léxica en memoria de Conchita Serrano (Manuales y Anejos de "Emerita,"* XLI), Madrid: National Foundation for Scientific Research, Center of Humanities, 1999, pp. 211–28.

A. Touwaide, *Loquantur ipsi ut velint ... dumtaxat non ignorent quis serpens sit tiro:* Leoniceno's contribution to the construction of Renaissance botanical lexicon, in *Medical Latin from the Middle Ages to the Early Modern Times: European Science Foundation. Proceedings of the Symposium held in Brussels, September 1999,* Leuven: Peeters-Orientaliste, 2000, pp. 151–73.

A. Touwaide, *Magna Graecia iterata:* Greek Medicine in Southern Italy in the 11th and 12th centuries, in Alfredo Musajo Somma, ed., *Medicina in Magna Graecia: The Roots of our Knowledge*, Bari: University of Bari, 2004, pp. 85–101.

A. Touwaide, Un manuscrit athonite du *Traité de matière médicale* de Dioscoride: l'Athous Magnae Laurae Ω 75, *Scriptorium* 45, 1991, pp. 122–27.

A. Touwaide, Medicina Bizantina e Araba alla Corte di Palermo, in Natale Gaspare De Santo and Guido Bellinghieri, eds., *Medicina, Scienza e Politica al Tempo di Federico II. Conferenza Internazionale, Castello Utveggio, Palermo, 4–5 ottobre 2007*, Naples: Istituto Italiano per gli Studi Filosofici, 2008, pp. 39–55.

A. Touwaide, *Medicinalia Arabo-Byzantina, Première partie: Manuscrits et texts*, Madrid: [the Author], 1997.

A. Touwaide, Le paradigme culturel et épistémologique grec dans la science arabe à la lumière de l'histoire de la matière médicale, *Revue du Monde Musulman et de la Méditerranée* 77–78, 1995, pp. 247–73.

A. Touwaide, Persistance de l'hellénisme à Bagdad au début du XIIIème siècle: Le manuscrit Ayasofia 3703 et la politique abbasside, *Erytheia* 18, 1997, pp. 49–74.

A. Touwaide, Pharmacology, in Thomas Glick, Steven J. Livesey, and Faith Wallis, eds., *Medieval Science, Technology, and Medicine: An Encyclopedia*, New York and London: Routledge, 2005, pp. 394–97.

A. Touwaide, Pharmacology, Pharmacy, in Albrecht Classen, ed., *Handbook of Medieval Studies: Terms, Methods, Trends*, 3 vols., Berlin and New York: Walter de Gruyter, 2010, pp. 1056–1090.

A. Touwaide, Pharmacy and materia medica, in Thomas Glick, Steven J. Livesey, and Faith Wallis, eds., *Medieval Science, Technology, and Medicine: An Encyclopedia*, New York and London: Routledge, 2005, pp. 397–99.

A. Touwaide, Pharmakologie (Frühmittalter, Hochmittelalter, Islamische Welt, Renaissance), in Hubert Cancik and Helmuth Schneider, eds., *Der neue Pauly. Enzyklopädie der klassischen Altertumswissenschaft*, vol. 15/2, Stuttgart and Weimar: Metzler Verlag, 2002, cols. 215–222. (English translation: Touwaide, Pharmacology, in Hubert Cancik and Helmuth Schneider, eds., *Brill's New Pauly Encyclopaedia of the Ancient World*, sup. vol. 4, Leiden and Boston: Brill, 2009, cols. 357–66).

A. Touwaide, Un recueil de pharmacologie du Xe siècle illustré au XIVe siècle: le *Vaticanus graecus* 284, *Scriptorium* 39, 1985, pp. 13–56.
A. Touwaide, Le strategie terapeutiche: i farmaci, in Mirko D. Grmek, ed., *Storia del pensiero medico occidentale*, vol. 1: *Antichità e Medio-Evo*, Rome and Bari: Laterza, 1993, pp. 353–73. (English translation: Therapeutic Strategies: Drugs, in Mirko D. Grmek, ed., *Western Medical Thought from Antiquity to the Middle Ages,* Cambridge: Harvard University Press, 1999, pp. 259–72 and 390–94).
A. Touwaide, Theoretical Concepts and Problems of Greek Pharmacology in Greek Arabic Medicine: Reception and Re-elaboration, *Forum* 6:1, 1996, pp. 21–39.
A. Touwaide, Tradition and innovation in Mediaeval Arabic Medicine: The translations and the heuristic role of the word, *Forum* 5:2, 1995, pp. 203–13.
A. Touwaide, La traduction arabe du *Traité de matière médicale* de Dioscoride: Etat de recherche bibliographique, *Ethnopharmacologia*, 18, 1996, pp. 16–41.
A. Touwaide, Transcultural Tradition of Pharmacological Data Through History: The Problems of the Illustration: A Case Study, in Christine E. Gottschalk-Batschkus and Christian Rätsch, eds., *Ethnotherapies: Therapeutic Concepts in Transcultural Comparison* (*Curare*, Special Issue), Berlin: Verlag für Wissenschaft und Bildung, 1998, pp. 175–78.
A. Touwaide, Transfer of Knowledge, in Albrecht Classen, ed., *Handbook of Medieval Studies: Terms, Methods, Trends*, 3 vols., Berlin and New York: Walter de Gruyter, 2010, pp. 1368–99.
W. Treatgold, *A History of the Byzantine State and Society*, Stanford: Stanford University Press, 1997.
M. Ullmann, *Die Medizin im Islam* (Handbuch der Orientalistik, Erste Abteilung, Der nahe und der mittlere Osten, Ergänzungsband, 6, Erster Abschnitt), Leiden: Brill, 1970.
G. Valla, *Giorgio Valla Placentino interprete in hov columine continentur Nicephorilogica ... Rhazes de pestilentia ...*, Venice: per Simonen Papiensem, 1498.
O. Volk, *Die byzantinischen Klosterbibliotheken von Konstantinopel, Thessalonike und Kleinasien*, PhD diss., Universität München, 1954.
N. Wilson, The Madrid Skylitzes, *Scrittura e civiltà* 2, 1978, pp. 209–19.

Keren Abbou Hershkovits

Galenism at the ʿAbbāsid Court

The first hundred years of the ʿAbbāsid Caliphate, the mid-eighth to mid-ninth century, are an interesting and fascinating period for many reasons. With regard to the history of medicine, it seems that these years were formative. It was a time of importing, translating, and appropriating knowledge. This in turn influenced the emergence of a medical corpus, the construction of the image of the good physician, and what it means to be one.[1] However, we have very few contemporary sources that describe this process of professionalization.[2]

One possible way to reconstruct it would be to establish a group of contemporary physicians, and then to draw a collective portrait.[3] However, in this paper I would like to put the identity of this group, and in particular their knowledge, into question; I wish to examine the forces that promoted physicians within the court. This paper therefore offers an alternative reading of the sources, and a different way of describing the reception of Galenism.

1 Introduction

Modern scholarship views the early centuries of Islam as a time of medical pluralism, and of a lively contest between different theories and physicians.[4] And yet, both the medieval and the modern narratives of the development of medicine single out Galenism, and the role of Christian physicians in promoting Galenism, and the translation of Greek texts into Arabic.[5] In what follows, I will present the modern narrative explaining the dominance of Galenic medicine in the Muslim world. Next, I will suggest a reconsideration of some of the as-

1 Dols, Origins of the Islamic Hospital, pp. 367–90; Pormann and Savage-Smith, *Medieval Islamic Medicine*.

2 I follow Paul Unschuld's definition of professionalization: "a process during which the control of a given group of health-care providers over resources pertaining to health care is increased" (Unschuld, The Physician in Imperial China, pp. 90–115).

3 Such a portrayal is available in Rosenthal, The Physician in Medieval Muslim Society, pp. 476–91.

4 Dols, review of *Health and Medicine*, p. 420; Pormann and Savage-Smith, *Medieval Islamic Medicine* (ch. 1).

5 Dols, *Medieval Islamic Medicine*, p. 6. See also ibid., p. 8: "The greatest impetus to medical studies in Islamic society came from the translation of Greek scientific works into Arabic."

sumptions of this narrative. Finally, I will propose some possible alternative explanations for this dominance.

2 The Modern Narrative: The Transmission of Medicine and the Dominance of Galenism

The narrative currently favoured by modern scholars presents medicine in Islam as based on the four humours theory and the Greek-Galenic medicinal system.[6] These reached Muslim scholars through a process of translation sponsored by various members of Islamic society including, perhaps even primarily, Nestorian-Christian physicians. These physicians were both practitioners and patrons of translations of Greek and Syriac texts into Arabic. Nonetheless, according to Savage-Smith and Pormann, Greek medicine was not the only system available. In early periods, i.e., the eighth to ninth centuries, other practices such as folk medicine, Persian, Indian, and Chinese medicine were in some use. However, due to the availability and dominance of Christian physicians following Galenic medicine, these other systems never generated enough support among courtiers (as either patrons or clients), while Galenism continued to grow stronger. As a consequence, medicine in medieval Islam is basically Galenic medicine. Though some remedies can be traced to different sources, still the great mass of concepts and remedies are Galenic.[7]

The dominance of Galenism in Islamic medicine has been explained by other factors as well. Dols argues that "The predominance of the Greek tradition was largely due to the Hellenized Christians, Jews, and Persians, who made up the bulk of the population in the newly established [Ummayad] empire and to the persistence of their centers of learning."[8] Hence, according to Dols, the preference and adherence to Galenism was not restricted to the elite or court circles only, but included the population at large. Galenic physicians were the healthcare providers of the regions before and after the Muslim conquest.[9]

---

6 Pormann and Savage-Smith, *Medieval Islamic Medicine*, pp. 23 – 29.

7 Pormann and Savage-Smith, *Medieval Islamic Medicine*, pp. 21: "[Indian and Chinese systems] also influenced the Islamic one, though only marginally. Through trade and travel, the Arabs came into direct contact with these two civilizations ...."

8 Dols, *Medieval Islamic Medicine*, p. 3.

9 The same argument is repeated by Pormann and Savage-Smith: "The immediate environment into which Islam erupted is the world of Late Antiquity where Greek thought pervaded not only Alexandria and other traditional strongholds of Hellenism, but also the Persian and Syriac

A third factor that allowed for the dominance of Galenism and promoted Greek medicine relates to a larger process of translation of scientific texts during the first two hundred years of the ʿAbbāsid caliphate. According to Dimitri Gutas, the ʿAbbāsid's interest and investment in translation was a part of their political agenda. In his *Greek Thought, Arabic Culture*, Gutas explains that the ʿAbbāsids wished to present themselves as the rightful heirs of the Sassanids, and therefore adopted Sassanid court etiquette.[10] One aspect of Sassanid court life was support for the translation of scientific texts into Pahlavi,[11] another was the preference for Christian physicians by Sassanid rulers.[12] The initiation of the translation movement by the second ʿAbbāsid caliph al-Manṣūr (r. 754–775) is therefore explained as part of his political agenda. The same reason is given for his inviting a Christian (and thus Galen-oriented) physician to attend to his medical needs. Both policies – support of translations and invitation of Christian physicians – continued under his successors.[13]

Modern scholarship identifies, then, two main reasons for the dominance of Galenic medicine. The first is practical: the availability of Galenic physicians and texts in the lands that composed Islamic territory. The second is political: the ʿAbbāsid's wish to continue Sassanid policies, hence the invitation and preference of Galen-oriented Christian physicians. According to this narrative, the dominance of Galenism was almost inevitable. In the next section, however, I will challenge this by problematizing these two key explanations for the adoption of Galenism.

---

speaking lands. [...] Islam adopted and adapted the scientific and medical heritage of those who came under its sway" (*Medieval Islamic Medicine*, p. 2).

10 Gutas, *Greek Thought, Arabic Culture*, pp. 34–45.

11 See ibid., pp. 25–26.

12 Jouanna reports that, until the time of Darius (550–486 BCE), the medical practitioners at the Achaemenid court were Egyptian. However, they failed to help Darius's dislocated ankle. After seven sleepless nights, he was informed that there was a physician among his Greek prisoners. He summoned him, and was quickly cured. In the years that followed – and much later, under the Sassanids – the emperors invited (or forced) Greek physicians to treat them, or members of their families. See Jouanna, *Hippocrates*, pp. 23–4.

13 For the invitation of the first Christian/Galenic physicians see Ibn al-Qifṭī, *Ikhbār al-ʿulamā'*, pp. 158–160; and Ibn Abī Uṣaybiʿa, *ʿUyūn al-anbā'*, vol. 1, pp. 123–125.

3 Reconsideration of Assumptions

3.1 Sources and the Problem of Sources

Information regarding medicine and physicians can, from the second half of the ninth century onward, be found in several early genres, among them professional medical texts, chronicles, and *belles lettres* literature (*adab*). However, most of our knowledge is derived from later sources – especially biographical dictionaries, some of which were dedicated to physicians. Among our sources is the *Fihrist* (catalogue) of Ibn al-Nadīm (a general biobibliographic dictionary written in tenth-century Baghdad),[14] and three biographical dictionaries discussing physicians.[15] The first of these latter appeared in late tenth century al-Andalus, and the others in twelfth to thirteenth century Damascus and Cairo. Together, they are: the *Kitāb ṭabaqāt al-aṭibbāʾ wa al-ḥukamāʾ* of Ibn Juljul (Andalusian, lived 944 to some time after 994 C.E.);[16] the *Ikhbār al-ʿulamāʾ bi-akhbār al-ḥukamāʾ* of al-Qifṭī (Aleppan, lived 1172–1248 C.E.);[17] and the *ʿUyūn al-anbāʾ fī ṭabaqāt al-aṭibbāʾ* of Ibn Abī Uṣaybiʿa (Cairene, lived 1203–1270 C.E.)[18] Additional information can be extracted from the wisdom literature, though this tends to focus more on sayings, rather than actual biographies.[19]

We do not have direct sources for the first encounters with healthcare providers or medical theories in the early period, and this lacuna has significant implications for information derived from later sources. The most obvious is that lack of first-hand contact with people and events necessitates a critical reading. Consequently, modern scholars take most of the information derived from biographical dictionaries with a grain of salt. Anecdotes are deconstructed and

---

14 Ed. Gustav Flügel, Leipzig, 1872.

15 For this unique Islamic genre, its character and structure, see: Bray, Literary Approaches, pp. 237–53; Gibb, Islamic Biographical Literature, pp. 54–8; Douglas, Controversy and its Effects, pp. 115–31; al-Qadi, Biographical Dictionaries: Inner Structure, pp. 93–122; *idem*, Biographical Dictionaries, pp. 23–75; Rosenthal, *History of Muslim Historiography*; and Young, Arabic Biographical Writing, pp. 168–87.

16 Ed. Fu'ād Sayyid, Cairo, 1955.

17 Ed. Julius Lippert, Leipzig, 1903.

18 Ed. Gerhard Müller, Frankfurt am Main, 1995 (reprint).

19 Wisdom Literature/aphoristic literature is discussed in detail in Gutas, *Greek Wisdom Literature* (see, esp., the introduction). This genre was designed for public consumption. The subject matter consists of the sayings of highly respected figures, though sometimes biographical information can be extracted from these texts.

sometimes proven false or wrongly attributed. A case in point is the myth of Jundishapur, which will be discussed below.[20]

However, a more significant issue is the interpretation and use of terms, ideas, and images present in the texts, i.e., the intellectual context. Islamic communities throughout the Muslim world underwent many changes and transformations during the concerned period. Most important for the present discussion is that, by the tenth century, medicine had become increasingly associated with Galenism. This association is evident in a large variety of genres, both medical and otherwise: ethical treatises, professional texts, discourses on medical education, hospital manuals, market inspection literature, *belles lettres*, chronicles, and biographical dictionaries. In all of these, the physician is a Galenic doctor, portrayed according to the Galenic criteria of a physician-philosopher.[21]

The descriptive choices (and, sometimes, the paucity of details) of the different authors reflect a long process of construction of what medicine is, and what "physician" stands for. Therefore, I would suggest setting aside the assumption that most physicians, or even most available physicians, were Nestorian and Galen-oriented; and, in what follows, I will ask what constituted the knowledge of physicians who visited the court in the early ʿAbbāsid period.

3.2 Was it Galenism they practiced?

The first physician commonly associated with Galenic medicine at the ʿAbbāsid court is Jūrjīs b. Jibrīl b. Bukhtīshūʿ. His education might shed some light on the prevailing medical knowledge (or medical training) of his time. According to Ibn al-Qifṭī:

> When al-Manṣūr founded *Madīnat al-Salām*, he had a stomach ache. [...] All treatment offered did not help him, and his illness got worse. He ordered al-Rabīʿ to gather the physicians. Once they gathered, al-Manṣūr told them: I want a good doctor from another city, an

20 See Ibn al-Qifṭī, *Ikhbār al-ʿulamāʾ*, pp. 133–34; and Pormann, Islamic Hospitals, p. 353. I wish to thank Prof. Pormann for allowing me to read the article before publication.

21 See Leiser, *Medical Education*, pp. 63–65; Pormann and Savage-Smith, *Medieval Islamic Medicine*, pp. 83–4; Savage-Smith, Ṭibb. As for hospital manuals and the use of Galenic medicine, see Pormann, Medical Methodology, pp. 95–118; and *idem*, The Physician and the Other, pp. 189–227. This image is clear in the writing of al-Jāḥīẓ. One of the heroes of his *Book of Misers* is Asad Ibn Jānī, a Muslim physician lamenting his lack of clientele. The reason, according to Asad, is his Islamic religion and Arabic origin; were he a Christian, he would have a large clientele (al-Jāḥiẓ, *Kitāb al-bukhalāʾ*, pp. 109–110; for R.B. Serjeant's English translation, see: *The Book of Misers*, pp. 86–87).

> expert (*māhir*). They answered that the best contemporary physician they knew was Jūrjīs b. Bukhtīshū', the head of physicians in Jundishapur. He was an expert in medicine and had significant (*jalīla*) treatises in medicine. [22]

Al-Manṣūr invited Jūrjīs, whose treatment proved successful, and Jūrjīs remained in the court as al-Manṣūr's doctor until 769.[23]

Not much information is disclosed in this short paragraph, though it seems that the reader is meant to appreciate Jūrjīs's qualifications from these few details. In fact, medieval authors associated Jundishapur with a medical school and a large hospital, as well as with translation activity. Greek and maybe Sanskrit medical texts were rendered there into Syriac and Pahlavi as early as the fourth century A.D. Therefore, Jūrjīs was invited not only because of his own reputation, but also because he represented a long tradition of learning and practice.

As indicated earlier, however, the association of Jundishapur with medical activity is considered by modern scholars to be a myth. Pormann and Savage-Smith conclude that Jundishapur might have had a small infirmary, as in neighbouring Susa, rather than a full-scale hospital. They suggest that information dispersed in medieval sources regarding Jundishapur is probably a late story disseminated by members of the Bukhtīshū' family, who, along with their associates, intended to justify their position as prominent doctors, and as conveyors of the best medical knowledge available.[24]

The debatable authenticity of this first encounter with a Galen-oriented Christian physician leads to other questions: What was Jūrjīs' medical knowledge? Was it in fact based on Galenic texts, and if so, which, and where did he find them? Were there many physicians with this same training and knowledge? One possible place to look for answers might have been the treatises Jūrjīs supposedly wrote; however, we have no information regarding these treatises, their contents, or purposes. There are no sources that quote them or provide an abridgement.

Another place for answers, though less favourable, might be the education of contemporary physicians in other regions. In the Byzantine Empire, according

---

22 Ibn al-Qifṭī, *Ikhbār al-ʿulamāʾ*, pp. 158 – 60. Ibn Abī Uṣaybiʿa repeats the same; see: *ʿUyūn al-anbāʾ*, vol. 1, pp. 123 – 4.

23 See Sourdel, Bu<u>kh</u>tī<u>sh</u>ūʿ.

24 One of the main problems is that Ibn al-Qifṭī himself was the first to write about Jundishapur as a medical center, several hundred years after the invitation of Jūrjīs (*Ikhbār al-ʿulamāʾ*, pp. 133 – 34). For a deconstruction of this narrative, see Pormann, Islamic Hospitals, p. 353; Pormann and Savage-Smith, *Medieval Islamic Medicine*, p. 21, n. 18; and Nutton, From Galen to Alexander, pp. 12 – 13.

to Vivian Nutton, it is not very clear what the nature of medical treatment or education was before the twelfth century. Though it is highly likely they were familiar with Greek medicine, there is no certainty as to what they knew and in what form.[25] As for the Latin West, it is not very clear, according to Peregrine Horden, that Galenic medicine was studied at that period for the purpose of engaging with medical practice. Horden shows that medical knowledge comprised many components, including Hippocratic and Galenic concepts, principles, and texts. However, he maintains that Galenic texts were not necessarily studied for their medical benefits. In centers such as Nisibis, it is more likely they were studied as part of a philosopher's education, but not for the kind of knowledge one would expect from an actual physician.[26] A similar picture arises from a study by Isabella Andorlini, who argues that in centers of scholarship such as Alexandria and Nisibis, Galen's texts were studied as part of a philosophical curriculum, rather than for medical reasons.[27] As for Anglo-Saxon England, Anne Van Arsdall demonstrates the same distinction between a practicing physician and a scholar who happens to study Galenic texts.[28]

At this point, it seems the assumption that most, or even a large number, of the healthcare providers in the Sassanid Empire relied on Galenic texts or guidelines would be naive. Nonetheless, Christian and Greek physicians were probably a part of the Sassanid court, though one cannot specify what their education might have been. Therefore the next question should be: Was the Sassanid court restricted to, or did it clearly prefer, Galenic/Greek/Christian physicians? In partial answer we find al-Thaʿālibī's (d. 1038) anecdote concerning physicians in the court of Shāpūr (r. 241–271). The Shah asked the king of India to send him a doctor, as he was unsatisfied with his Greek physicians, so the Indian king sent him a physician whose medical knowledge was provided by revelation. Thus, in this case, the physician's personal charisma, rather than his theoretical knowledge, was of importance. In addition, al-Thaʿālibī states that, from the time of Shāpūr, the inhabitants of Susa were knowledgeable in medicine; they had an Indian physician and Greek prisoners of war to teach them.[29] These two remarks appear to put Indian and Greek physicians on the same level, perhaps even in contest.

---

25 Nutton, *Ancient Medicine*, pp. 157–59.

26 See Horden, What's Wrong with Early Medieval Medicine?; and *idem*, How Medicalised Were Byzantine Hospitals?

27 Andorlini, Teaching Medicine in Late Antiquity, pp. 401–14, esp. at 406.

28 Van Arsdall, Medical Training in Anglo-Saxon England, pp. 415–43, esp. at 427.

29 al-Thaʿālibī, *Histoire des rois des Perses*, pp. 531–32; Dols, Origins of the Islamic Hospital, p. 378.

Other instances of exchanges of knowledge between India and Persia before Islam are recorded. Khosraw Anūshirwān (r. 532–578) sent his personal physician Borzawayh to India, in order to bring back works of science and medical knowledge and have them translated.[30] One may assume the Shah is here demonstrating an interest in, and perhaps an appreciation of, Indian medical knowledge.

There is also evidence of philosophical encounters with Greek philosophy, through Christian scholars and churchmen. One example is Paul the Persian, who dedicated to Khosraw a book, in Syriac, on Aristotelian logic.[31] In addition there is evidence of Aristotelian influence in the *Dēnkard* (an apologetic encyclopedia of Zoroastrianism).[32] However, it is not clear if this influence was longstanding or new. According to Bailey, the *Dēnkard* was finalized during the ninth century; it is difficult to determine if these Aristotelian influences are ninth-century additions, or reflective of earlier traditions.[33] This obscurity led Bailey to argue: "In the absence of adequate contemporary evidence as well as the chronological and linguistic problems in the extant Zoroastrian Pahlavi sources, which largely date from the ninth century, it is not possible to establish the extent of Greek influences and learning under the Sassanids."[34]

Moreover, we do not have conclusive contemporary information regarding the knowledge of physicians, their education, and – more importantly – whether or not they followed a particular theoretical framework. There is no clear indication as to what practicing physicians, including Jūrjīs b. Bukhtīshūʿ, knew, what was available to them, or even what they read. Medical theory and practice in pre-Islamic Iran are a chapter in world history that has yet to be written.[35] For

30 Dols, Origins of the Islamic Hospital, p. 381; Meyerhof, Transmission, pp. 24–29.
31 Gutas, Paul the Persian, pp. 231–67.
32 Russell, Christianity in Pre-Islamic Persia.
33 Bailey, *Zoroastrian Problems*, p. 79.
34 Ibid., p. 81.
35 The only available textbook treating Iranian medicine before the Islamic conquest is Cyril Elgood's *A Medical History of Persia and the Eastern Caliphate from the Earliest Times until the year A.D. 1932*. However, this book is outdated and some of the data is considered inaccurate. A more recent survey of medicine in pre-Islamic Iran appears in the *Encyclopaedia Iranica:* see Gül, Greek Medicine in Persia. However, the article reflects similar difficulties in assessing the education and availability of Greek medical texts under the Sassanids. The question of medicine and religion in Iran also has a modern twist. A certain attitude toward modern (Western) medicine in Iran (and also in India) considers it part of a colonial attempt at control. This conception has led some to return to Galenic medicine, or *Yunānī Ṭibb*, which is considered part of the prophetic legacy of Idrīs and, later, Muḥammad. See Ebrahimnajed, The Development of Galenic-Islamic Medicine, pp. 127–40.

now, it would be safer to assume that the Sassanid Shahs did not confine themselves to a single medical system.[36]

In light of the above, the introduction of Galenism to the Muslim world cannot be explained only by a direct continuation of Sassanid policy, as there is no indication that such a policy actually existed. Likewise, it is not clear that Galenism prevailed among the general population.[37] Indeed, the early days of ʿAbbāsid rulership demonstrate an eclectic attitude, which does not comply with the "Galenism as an inevitable option" narrative. Galenism was not the only healthcare option available to royalty or to ordinary people;[38] and for royalty at least, there is sufficient evidence that it was not even the most prevalent. The very same al-Manṣūr who is credited with the invitation of the first Christian/Galenic physician, also used the services of an Indian physician when his current doctors could not suggest an appropriate solution.[39]

Biographical dictionaries and chronicles record the visits and invitations of various physicians to the court under various circumstances. Physicians were usually invited on account of their reputation, rather than doctrinal affiliation. In some cases, the sources mention the professional knowledge these physicians had, while in other cases they attribute to them a lack of knowledge. Consequently, I would like to suggest a different angle for studying the transmission of medicine. Instead of defining a rule – i.e., that Christian physicians were the main outlet of medical knowledge and held high influence in the court – let us assume that there was no rule, and therefore no exception to the rule, at least between the mid-eighth and mid-ninth centuries. Let us now return to the sources and read their stories.

---

36 According to Pingree, the astronomical and astrological traditions of the Sassanids demonstrate different influences, including Indian and Greek, of different periods. See Pingree, Greek Influence, pp. 32–43.

37 Moreover, as suggested by Guy Attewell (Islamic Medicines, p. 334), even if the Sassanid elite, and/or population at large, had preferred Galenism to any other kind of medicine, this would not explain why the ʿAbbāsids did. Why would one assume that the ʿAbbāsids actually wished to engage with the local Sassanid court-cultures? In the same vein, it is not obvious that Greek sciences would have been accepted either.

38 Peter Pormann's study regarding the "other" in medicine (The Physician and the Other, pp. 189–227) mentions various practitioners that Galenic physicians often criticized; this constitutes evidence that different practices actually existed and were prominent at particular times and places, or among particular groups.

39 See al-Ṭabarī, *Taʾrīkh al-rusul wa al-mulūk*, vol. 10, pp. 387–8. For Hugh Kennedy's English translation, see *idem*, *The History of al-Ṭabarī*, vol. 29, p. 89. Interestingly, this information does not appear in biographical dictionaries dedicated to physicians.

Physicians were invited to court – and sent away from it – on a regular basis. Invitations were sometimes a demonstration of faith in the new physician (as with Jūrjīs b. Bukhtīshū'), but in many cases, they were rather a demonstration of dissatisfaction with the ones already there (as with the Indian physician invited to al-Manṣūr's sickbed). Reasons for dismissal varied, ranging from accusations of incompetence (as with the physicians of al-Hādī)[40] to accusations of disloyalty (as with Ibn Shahlāfā).[41]

Physicians were invited to the court because the caliph wanted a solution for a situation, either his own sickness or that of a loved one. This might imply that the theoretical adherence of the physician was not the most important motivation for the invitation/sending-away of a health care provider; the situation and the people it concerned were more important. It could also reflect upon the role (or high standing) of the person who suggested the name of the new doctor; i.e., importance lay with the inviter rather than the invited. Whatever the case, the diversity of knowledge and character of each physician should not necessarily be read as competition between theoretical principles.

Nevertheless, this perspective does not explain the strong position eventually achieved by Christian/Galenic physicians. The question remains: What happened – what were the motivations that allowed them to gain dominance? What turned Galenism into the prevailing medical system and helped to reject other available healthcare options? In the next section I will suggest a few possibilities.

---

40 When the caliph al-Hādī (d. 786) was on his deathbed, he was very disappointed with the medical care he received from his physicians, arguing they were only taking his money and gifts but not treating his illness. He wished Bukhtīshū' to come and treat him; he was summoned, but did not make it in time. Al-Rabī' informed the Caliph that he knew of a Persian physician from the area of the Ṣarṣar River, and this physician was invited to treat al-Hādī. There is no information regarding the medical training he had, or his affiliation to a particular medical school. Nor, in fact, did he truly offer any medical treatment. Rather, this physician told the Caliph he knew the treatment, asked for the ingredients and help he needed to prepare the medication, but told everybody else the Caliph was about to die (Ibn Abī Uṣaybi'a, *'Uyūn al-anbā'*, vol. 1, p. 126).
41 'Īsā b. Shahlāfā was dismissed from his post as a court physician, and his property confiscated, in the time of al-Manṣūr. The reason was his blackmailing of the Nestorian Patriarch, and his bragging that the life of the caliph was in his hands (Ibn al-Qifṭī, *Ikhbār al-'ulamā'*, p. 248; Ibn Abī Uṣaybi'a, *'Uyūn al-anbā'*, vol. 1, p. 135).

4 Why Pluralism Disappeared: Political Aspects and Patronage

The translation of Galenic texts was not a simple process of taking up readily-available works and drawing upon accessible practices. It was a consuming process that demanded many resources and much support; texts needed to be found, and translating skills and a proper vocabulary had to be developed. Patrons had to pay for texts to be brought from different places, translated, and commented upon. The dominance of Galenism, which is evident in the tenth century, is the outcome of different aspects of this process, but cannot explain its details or the actual choices involved.

It has been suggested that the involvement of Christians in the translation movement in general allowed for translators to promote particular text-choices. Modern scholars emphasise the part played by Ḥunayn b. Isḥāq (808–873) and other Christian scholars in the translation movement. Ḥunayn is usually credited with creating a scientific vocabulary in Arabic and a most efficient translation method.[42] A further study into the role of Christians as translators, however, leads to reading ecclesiastic texts, and in particular a series of letters written by Timothy the Patriarch.[43] These letters are a rich lode of information regarding the Christian community, the clergy, and the interaction between them and the caliphs.

Several of Timothy's letters refer to translations he made for the caliph al-Mahdī (r. 775–785). Of significant importance is his letter (number 43) to Rabban Mar Pethion, the head of the school of Mar Abraham, in which he requests that Mar Pethion find some needed books. In this letter he tells Mar Pethion that he translated Aristotle's *Topics* for al-Mahdī, and also inquires whether Mar Pethion knows of other translations of Aristotle's texts. It is evident that his was not the only translation available to the caliph, as Timothy mentions another; however, whereas Timothy's translation was made from the Syriac, the other was made from the Greek. Both translations were available to the caliph, and he read them both. Nonetheless, Timothy states that al-Mahdī preferred his translation, rather than the one translated from Greek.[44]

---

42 For Ḥunayn's achievements, see: Strohmaier, Ḥunayn b. Isḥāḳ al-ʿIbādī. See also: Meyerhof, New light on Ḥunain Ibn Isḥâq, pp. 685–724.

43 I wish to thank Prof. Dimitri Gutas for referring me to these letters, which were invaluable for researching the present paper.

44 The translation mentioned in Letter 43 is not extant. Al-Nadīm, the aforementioned tenth-century biobibliographer, does not list a translation made by Timothy when discussing the avail-

Brock's analysis of this letter explains Timothy's request as part of a competition between translators and translation methods; Timothy was anxious to be able to supply the best translations, and to get hold of texts that might be of interest to the caliph.[45] This competition also had a political aspect to it: the Nestorians were suspected of disloyalty, of secretly resenting the Islamic rulers and spying for the Byzantines.[46] Timothy thus used translation as a door to the court, to prove his loyalty and to demonstrate the service he was, uniquely, able to provide.

These translations, however, were of philosophical texts; the first specifically medical translations of which we have concrete evidence date from the middle of the ninth century: those of Ḥunayn and of Yuḥannā b. Māsawayh (d. 857).[47] Earlier translations of medical texts were probably made from Sanskrit or from Persian dialects. Several names of Indian physicians and translators appear in al-Nadīm's *Fihrist*,[48] and further evidence of Indian texts and exchanges can be derived from ʿAlī b. Rabbān al-Ṭabarī's *Firdaws al-Ḥikma* (Paradise of Wisdom), written in the middle of the ninth century. This work intended to provide the reader with a general introduction to medicine; the author found Indian medicine important enough to dedicate a chapter to it, explaining with precision its basic elements. The information appearing in this chapter demonstrates that, at the time of writing, ʿAlī b. Rabbān had sufficient available information, and probably enough interaction with Indian physicians and scholars, to enable an accurate summary.[49] The involvement of Indian physicians in early ʿAbbāsid times is a subject that deserves further inquiry and study, and I hope to pursue this line of research in the near future.[50]

Returning to Timothy, the translations he made for the caliph were not his only link to the court: he also participated in its religious debates and was much respected by its different members. He was also supported by Nestorian members of the court; namely, the physicians ʿĪsā Abū Quraysh and Jibrīl b.

---

able translations of the *Topics*. He does refer to an ancient translation, but provides no details with regard to it (*Fihrist*, p. 249).

45 Brock, Two Letters, pp. 233–46. A short summary of the letter also appears in Bidawid, *Les Lettres du patriarche*, p. 35. The letter was written ca. 782 (ibid., 66).

46 See Fiey, *Chrétien Syriaque*, p. 37.

47 Dallal, *Islam, Science*, p. 31. For Yuḥannā, see Vadet, Ibn Māsawayh.

48 *Fihrist*, pp. 271, 301.

49 *Firdaws al-Ḥikmah, pp. 557–594.* See also: Meyerhof, ʿAlī at-Tabarī's "Paradise of Wisdom," pp. 6–54; and Khan, Ali Ibn Rabban aṭ-Ṭabari, pp. 1–14.

50 One recent study, discussing the different spheres of interaction between early the Islamic Caliphate and India, as well as later periods, is Finbarr B. Flood's *Objects of Translation: Material Culture and Medieval "Hindu-Muslim" Encounter.*

Bukhtīshūʿ. Both were involved in various inter-Christian developments, and their names are mentioned in relation to the appointments of patriarchs.[51] How did they reach a position that allowed them to influence the whole Christian community?

Medieval sources discuss the involvement of physicians in non-medical issues and their role as intermediaries or messengers; physicians were more than healers; they were courtiers, companions and prominent figures in their home communities.[52] Like other courtiers, they formed alliances and animosities. Doctors were part of a social network within the court, and these networks allowed for the inclusion or exclusion of a particular person from court and professional life. Someone from within might suggest their names and potential as physicians, or drop hints of disloyalty to send them away. This was a complex network that included home communities (usually the Christian community and the clergy), the professional milieu, and other courtiers, viziers, chamberlains, *qāḍī*s, and also, as will be discussed below, women of the harem. The power of physicians was directed to various facets of their lives, with some aiming to affect the policies of their home communities (nominations of local rulers, patriarchs, and mediation between court and community). Part of their power was exercised toward fellow physicians, creating categories for admission to the group, who could be defined as a doctor, and how to exclude those who did not fit in.[53]

According to Micheau, it was not political but private patronage that played the crucial role in the promotion of Galenic texts. He points to two main groups who supported translations. First are Christian physicians, from the days of al-Hādī to the time of al-Mutawakkil. Such were part of a network of families:

---

51 The influence and involvement of these physicians in court affairs are evident in both Christian and Muslim sources. Hārūn al-Rashīd, for instance, is reported to have said the following: "Any person that wants something from me, should ask Jibrīl [to make the request] as I do everything he asks me to" (Ibn al-Qifṭī, *Ikhbār al-ʿulamāʾ*, p. 135; Ibn Abī Uṣaybiʿa, *ʿUyūn al-anbāʾ*, vol. 1, p. 127); whereas Timothy the Patriarch defined Jibrīl b. Bukhtīshūʿ as: "Our door to our glorious king" (see Bidawid, *Les Lettres du patriarche*, pp. 77–8). Bodydena Wilson's 1974 PhD dissertation, *The Bakhtishuʿ: Their Political and Social Role under the ʿAbbasid Caliphs (A.D. 750–1100)*, is dedicated to the role played by different members of the Bukhtīshūʿ family in court and ecclesial affairs.

52 Physicians as courtiers are widely studied. For examples of the different roles played by doctors see Nutton, From Galen to Alexander, p. 13.

53 Ḥunayn's autobiography is a case in point. He complains that he is not appreciated by his colleagues, and not recognized as a physician. On two occasions he is cast out of the medical milieu, but both incidents end with his triumph and an acknowledgement of his skills and medical knowledge. See Cooperson, Ḥunayn ibn Isḥāq, pp. 107–118.

the Bukhtīshū‘, the Māsawayh, the Sarābiyūn, and the Tayfūrī. They lived in the same neighbourhood, married among themselves, and exchanged knowledge; but they also took part in intrigues and animosities. They were no longer comfortable with Greek, but needed the texts as a means of associating themselves with a particular body of knowledge. Their possessing a text unknown to others was a means for arguing a unique and better knowledge of medicine than that of their colleagues. The second of Micheau's groups is one of Muslim notables. Scholars such as al-Kindī and the Banū Mūsā supported translation as part of a wish to define their social status. Supporting translations was part of the consolidation of a new social elite; and so, Micheau concludes, the translation and support of translation of Galenic texts was not directly related to professional issues or medical preferences, but to a social context.[54]

The question of patronage and its influence on the promotion or inhibition of medical systems is not completely clear yet. It is not clear, for instance, why patrons of non-Galenic systems never generated enough support for their choice of medical assistance; and I would like to highlight a related aspect in need of elucidation: women's patronage. Abū Quraysh's ticket to the court was Khayzurān, al-Mahdī's wife and a prominent figure at the court.[55] Jibrīl b. Bukhtīshū‘

---

54 Micheau, Mécènes et médecins, pp. 176–9. Dols already recognized this division (Origins of the Islamic Hospital, p. 385). However, he saw the physicians' involvement as purely professional, whereas Micheau points out the social and political advantages attached to the support of translations. For more information on the social factor of the translation movement, see Gutas, *Greek Thought, Arabic Culture*. As for the animosity between al-Kindī and the Banū Mūsā, see Endress, The Circle of al-Kindi, pp. 43–76.

55 The relation between Khayzurān and her physician needs some explaining. According Ibn al-Qifṭī and Ibn Abī Uṣaybi‘a, when Khayzurān suspected she was pregnant, she sent her maid to give a urine sample to the court physician, al-Tayfūrī. While on her way, a person stopped her and asked: "To whom does this urine belong?" The maid answered that the urine was of a weak woman. He continued: "No, [this urine belongs to] a great queen, who is pregnant with a king." The maid went to complete her task; when arriving back at the harem she informed her lady that the answer she received from al-Tayfūrī confirmed the pregnancy, but he could not state the sex of the fetus. The maid also mentioned the strange incident she had had. Khayzurān kept the information close; and she indeed gave birth to a baby boy, the future caliph al-Hādī. She then informed her husband of the physician who had such wonderful skills and could foresee what al-Tayfūrī, the caliph's own physician, could not. She asked for that person to be brought to the court, bestowed presents upon him, and gave him the epithet Abū Quraysh, so everyone might understand his importance for the entire Islamic community. The last part of the anecdote relating the circumstances that led to Abū Quraysh's arrival at the court reports a test administered to him by al-Mahdī himself. He asked some medical question, but found that "he only knew very little pharmacology." Nevertheless, Abū Quraysh gained a respectable place in the caliph's entourage. (Ibn al-Qifṭī, *Ikhbār al-‘ulamā’*, pp. 430–31; Ibn Abī Uṣaybi‘a, *‘Uyūn al-anbā’*, vol. 1, pp. 149).

served another strong ʿAbbāsid lady, Zubayda, the wife of Hārūn al-Rashīd and the mother of al-Amīn (r. 809 – 813).[56] Among the people who paid him for services one also finds al-ʿAbbāsa, Hārūn al-Rashīd's sister, and Fāṭima Umm Muḥammad.[57] I would like to suggest that the patronage given to particular physicians by ladies of the ʿAbbāsid family had some influence on the course of history.

Both Khayzurān and Zubayda are mentioned as patrons of a particular physician, and as being attended by a specific physician. The relationship between Khayzurān and Abū Quraysh were probably more than mere occasional medical consultations. Not only was she responsible for his presence in court, she also maintained his high position there; when competition appeared she saw to its removal.[58] This high position at court allowed Abū Quraysh to play a significant role in the Christian community and to act as a middleman between the court and the patriarchate.[59] Zubayda's name is associated with two physicians, both Christians. Dāwūd b. Sarābiyūn al-Mutaṭabbib was appointed as her personal physician by Hārūn al-Rashīd;[60] and, in addition, there was a special connection with Jibrīl b. Bukhtīshūʿ. Fiey suggests that the nature of these relations was due to Zubayda's general sympathy to the Christian community; she gave presents to Timothy, sponsored Christian celebrations and the manufac-

56 Evidence of his strong position at court is the story told regarding the reasons that led Jūrjīs to leave. According to Ibn al-Qifṭī (*Ikhbār al-ʿulamāʾ*, pp. 160 – 1), Jūrjīs was not feeling well, so he asked to be dismissed and return to his home and family. However, according to Ibn Abī Uṣaybiʿa (*ʿUyūn al-anbāʾ*, vol. 1, p. 126), al-Manṣūr allowed Jūrjīs to return home, since he saw the hate expressed by Abū Quraysh and Khayzurān against him. Ibn Abī Uṣaybiʿa provides other examples for the animosity between them. When Jūrjīs learned that Abū Quraysh had told Khayzurān that she was pregnant with a boy, he called Abū Quraysh a liar. Once Khayzurān became pregnant again, he suggested Abū Quraysh once again try his talent, and thus show his fraud. But once again Abū Quraysh gave the correct answer and foresaw the birth of a male child. According to Ibn Abī Uṣaybiʿa, Abū Quraysh admitted to Jūrjīs that he simply had a lucky guess (twice) and not any special talent; thus Jūrjīs found it hard to see Abū Quraysh as his peer, and certainly not his superior in court (Ibn Abī Uṣaybiʿa, *ʿUyūn al-anbāʾ*, vol. 1, pp. 149).

57 Ibn Abī Uṣaybiʿa, *ʿUyūn al-anbāʾ*, vol. 1, pp. 136. I could not make a definite identification of Fāṭima Umm Muḥammad. The only Fāṭima that fits time and place is Fāṭima the daughter of al-Rashīd (see al-Masʿūdī, *Murūj al-dhahab*, vol. 6, p. 285).

58 Ibn al-Qifṭī, *Ikhbār al-ʿulamāʾ*, pp. 101, 431.

59 For the role played by Abū Quraysh as a middleman, and his involvement in ecclesial affairs, see Bidawid, *Les Lettres du patriarche*, p. 4; and Fiey, *Chrétien syriaque*, pp. 32 – 3, 36 – 7, 41, 55 – 6.

60 Ibn al-Qifṭī, *Ikhbār al-ʿulamāʾ*, pp. 438, 431.

ture of gold and silver crosses, and also helped Jibrīl b. Bukhtīshū' acquire permission for the reconstruction of a church in Baṣra.[61]

Zubayda's close ties with Jibrīl b. Bukhtīshū' brings to the fore the possibility that female patronage of particular physicians had some part in promoting Christian physicians in general, and therefore Galenic medicine. Women provided yet another source of strong patronage, a source of prestige and access to a somewhat closed realm. This kind of patronage might have had additional, less visible, importance; did the action of these ladies have any impact on the general public in terms of fashion? Did other women, striving to resemble the important ladies of the time, copy their preference in medical care, as they copied their taste in clothing? In his introduction to *Women in the Medieval Islamic World: Power, Patronage and Piety*, Gavin Hambly surveys available studies regarding female involvement in court activity.[62] These women, usually mothers and wives of caliphs, exercised great influence on various matters in the courts of their husbands and sons; some even set the political tone.[63]

The involvement of 'Abbāsid women in scholarly activity has not been studied at all in the 'Abbāsid context;[64] not only have scholars not been very interested in such information (it does not fit the fantastic world of sexual permissibility associated with the harem),[65] but the sources are not very generous in the information they disclose. The few studies available regarding women and medicine usually relate to two aspects. The first is women as providers of medical care: the study of their access to medical knowledge and of incidents where they are mentioned as practitioners.[66] The second aspect studied by modern

---

61 For Zubayda's acts of charity see al-Mas'ūdī, *Murūj al-dhahab*, vol. 8, p. 295. See also Fiey, *Chrétien Syriaque*, pp. 57 – 9; and Abbott, *Two Queens of Baghdad*, p. 240. For her close relations with Timothy the Patriarch, see Bidawid, *Les Lettres du patriarche*, p. 77.

62 Hambly, Becoming Visible, pp. 3 – 27.

63 The most prominent study of the role of women in the early 'Abbāsid period is still Nabia Abbott's *Two queens of Baghdad: Mother and Wife of Hārūn al-Rashīd.* A more recent study is Bray, Men, Women and Slaves, pp. 121 – 46.

64 For the Andalusian context, however, there are several studies regarding females educated in sciences other than the religious. See Ávila, La Mujeres 'Sabias' en al-Andalus, p. 166; and Marín, *Mujeseres en al-Ándalus*, pp. 653 – 6. I thank Dr. Cristina Álvarez Millán for bringing these fascinating studies to my attention.

65 For a description of the tenth-century harem, and a critical survey of the way modern scholars depict and relate to the harem in different regions, see El Cheikh, Revisiting the Abbasid Harems, pp. 1 – 19.

66 See Pormann, The Art of Medicine, pp. 1598 – 9; Shatzmiller, Aspects of Women's Participation, pp. 36 – 58 (esp. p. 42 for midwives in Geniza documents from the Mamlūk period); and Pormann and Savage-Smith, *Medieval Islamic Medicine*, pp. 103 – 5. See also: Giladi, *Muslim Midwives.*

scholars is women as patients. Miri Shefer's study is an examination of possible avenues to medical care for women in the Ottoman harem.[67] Ladies of the harem probably needed physicians, but we do not know if they had real access to them in the late eighth century. It is possible that they never saw the physician personally, and yet they could exercise great power on the course of things by simply suggesting the invitation of a particular doctor (as done by Khayzurān), the consultation of one, or the rejection of the advice of another. Incidents of women asking or paying for healthcare options might be seen as personal interests with little influence on the course of events, but must this really be the case? It is possible the support these physicians received reinforced their position in the court and allowed them to introduce other members of their family. It is also possible that other women looked to these prominent ladies for models and behavioral preferences.

Other women were involved in other aspects of healthcare, with requests for medical texts being particularly worthy of consideration. The first woman recorded as making such requests was the mother of al-Mutawakkil; in the biography of Ḥunayn b. Isḥāq, Ibn al-Qifṭī mentions a treatise written for her by Ḥunayn: the *Kitab al-Mawlūdīn.*[68]

A final episode of women's involvement in medicine that I will discuss here involves Shaghab, the mother of al-Muqtadir (r. 908–32); to her is attributed the founding, in 918 A.D., of a hospital in Baghdad, and the endowment of money for the healthcare of pilgrims. According to Pormann and Savage-Smith, these are not only important as acts of charity; they demonstrate involvement and interest in public health, as hospitals "improved the medical care available to the wider community."[69] According to El Cheikh, Shaghab's actions should be considered an illustration of female charity, building on the examples of Khayzurān and Zubayda. Furthermore, Shaghab's establishment of this particular hospital set in motion a long tradition of rulers establishing hospitals in most regions

67 See Shefer-Mossensohn, Sick Sultana, pp. 281–312, for a discussion of the healthcare options of elite women in the Ottoman world. There is little information regarding access to medical care for ʿAbbāsid women, though there is evidence of payment to physicians by women, and appointments of physicians to women in the caliph's household (see above). There are also comments regarding the permission physicians received to treat women. In Bukhtīshūʿ's biography, for instance, it is said that the caliphs trusted him with the mothers of their sons (Ibn al-Qifṭī, *Ikhbār al-ʿulamāʾ*, p. 101).

68 Ibn al-Qifṭī, *Ikhbār al-ʿulamāʾ*, p. 173. For women's ownership and authorship of medical texts in sixteenth-century Europe, see Green, *Making Women's Medicine Masculine*, pp. 304–9.

69 See Pormann and Savage-Smith, *Medieval Islamic Medicine*, p. 105.

of the Muslim world.[70] Although there is some evidence of two other hospitals prior to this, established by the Barmakids and by Hārūn al-Rashīd at the end of the eighth century, there is no information regarding the services they provided or for how long they operated.[71]

5 Summary

According to Attewell and Conrad, the lack of an institutionalized licensing process demanded other methods of legitimization.[72] Physicians needed some kind of tool to decide who could be a part of – and who should be excluded from – the group of legitimate physicians. One prominent method was the writing of treatises. These texts served as a demonstration for other physicians or possible patrons of the author's knowledge, as an exhibition of proficiency and eloquence – like business cards or a university diploma. The written word had great importance, and excluded all those who did not express themselves by writing in Arabic. Therefore, having no patrons meant not only having no access to court, or to posts of power, but also no funding for writing and translations; this in turn meant less recognition by other physicians and patrons, and fewer means to display skill, knowledge, and prestige.[73] In order to better understand

---

70 For al-Muqtadir's Greek slave mother Shaghab (her name means "troublesome"), and her involvement in court politics, see El Cheikh, Revisiting the Abbasid Harems, pp. 3, 9–12. Shaghab is important not only for her great political influence (some say she was the actual ruler, was involved in the appointment of a vizier, and so on), but also for her personal wealth, which she used, among other things, for medical activities (endowing, as we have seen, money for the provision of medical assistance to pilgrims). This should be viewed as a continuation of the practises of previous influential women, particularly Khayzurān and Zubayda. Zubayda in particular is known for her endowments for pilgrims and pilgrimage. For Shaghab's sending physicians to treat pilgrims, see Ibn al-Jawzī, *Muntaẓam*, vol. 13, p. 321; and El Cheikh, Gender and Politics, pp. 144–55. For Khayzurān's and Zubayda's acts of charity, see Abbott, *Two Queens of Baghdad*, pp. 124–5, 237–47. As for Hārūn al-Rashīd and Zubayda being role models for others to follow in leisure and in clothing, see al-Mas'ūdī, *Murūj al-dhahab*, vol. 8, pp. 295–6 (English translation: al-Mas'ūdī, *The meadows of gold*, pp. 387–392). For the contribution of 'Abbāsid ladies to medicine and medical practice, see: Abbou Hershkovits, Medical Services and Abbasid Ladies, pp. 121–136.

71 For an analysis of the establishment of hospitals during the reign of al-Muqtadir, see Pormann, Islamic Hospitals, pp. 337–82.

72 Attewell, Islamic Medicines, p. 343; and Conrad, Scholarship and Social Context, pp. 84–100.

73 See Totelin, *Hippocratic Recipes*, pp. 95–97.

the transmission of Galenic medicine, one needs first to unpack the networks of patrons and physicians.

In this paper, I have attempted to raise some issues relating to the way Galenism gained dominance. I have suggested that Galenism triumphed because a particular group of doctors (out of several) managed to fix the knowledge of Galenic texts as the fundamental criterion for admission to their ranks, and, therefore, a prerequisite for access to the court. This triumph was not an inevitable outcome, since Galenism had rivals; however, these rivals were (one way or another) eliminated.

This paper also demonstrates several problems pertaining to the prevailing narrative relating to the reception of the Galenic medical system. I have shown that the reception of Galenism was not an inevitable process; in fact, it was a slow process that took place over a long period of time, and, at any given moment, could have taken a different path. Nor can the transmission of medical knowledge be separated from other aspects of the time: the Nestorians' struggle for recognition as a valuable part of Islamic society, the sudden turn away from Buddhist knowledge, and the consolidation of cultural categories for what constitutes illness, health, and man's obligation with regard to his body. Political aspects proved important as well. The foundation of hospitals illustrates a conception that rulers were, in a way, obligated to their subjects' health – should this obligation be considered religious or political?

All of these aspects, and many others, are intertwined with one another, and reflect a cultural context which needs to be unearthed and studied from a new perspective; mainly, that Galenism was something people fought over – it was not given for free. The process that allowed for the dominance of Galenism reflects other social, political, and religious developments, and should be studied in their light.

Bibliography

N. Abbott, *Two Queens of Baghdad: Mother and Wife of Hārūn al-Rashīd*, London: Al Saqi, 1986.

K. Abbou Hershkovits, Medical Services and Abbasid Ladies, *Hawwa* 12/1, 2014, pp. 121–136.

I. Andorlini, Teaching Medicine in Late Antiquity: Texts and Contexts, in Patrizia Lendinara, Loredana Lazzari, and M.A. D'Aronco, eds., *Form and Content of Instruction in Anglo-Saxon England,* Turnhout: Brepols, 2007, pp. 401–14.

G. Attewell, Islamic Medicines: Perspectives on the Greek Legacy in the History of Islamic Medical Traditions in West Asia, in Helaine Selin, ed., *Medicine across Cultures: History and Practice in Non-Western Cultures,* Dordrecht, Boston: Kluwer Academic Publishers, 2003, pp. 325–50.

M.L. Ávila, La Mujeres 'Sabias' en al-Andalus, in María, J. Viguera, ed., *La mujer en al-Andalus: Reflejos históricos de su actividad y categorías sociales,* Madrid: Universidad Autónoma de Madrid, 1989, pp. 139–84.

H.W. Bailey, *Zoroastrian Problems in the Ninth-Century Books,* Ratanbai Katrak Lectures, Oxford: Clarendon, 1943.

R. Bidawid, *Les Lettres du patriarche Nestorian Timothée I,* Vatican City: Biblioteca Apostelica Vaticana, 1956.

J. Bray, Literary Approaches to Medieval and Early Modern Arabic Biography, *Journal of the Royal Asiatic Society, Series 3,* 20/3, 2010, pp. 237–53.

J. Bray, Men, Women and Slaves in Abbasid Society, in Leslie Brubaker and Julia M.H. Smith, eds., *Gender in the Early Medieval World East and West, 300–900,* Cambridge: Cambridge University Press, 2004, pp. 121–46.

S. Brock, Two Letters of the Patriarch Timothy from the Late Eighth Century on Translations from Greek, *Arabic Science and Philosophy* 9, 1999, pp. 233–46.

L.I. Conrad, Scholarship and Social Context: a Medical Case from the Eleventh-Century Near East, in Don Bates, ed., *Knowledge and the Scholarly Medical Tradition,* Cambridge: Cambridge University Press, 1995, pp. 84–100.

M. Cooperson, Ḥunayn ibn Isḥāq, in Dwight F. Reynolds, ed., *Interpreting the Self: Autobiography in the Arabic Literary tradition,* Berkeley: University of California Press, 2001, pp. 107–118.

A.S. Dallal, *Islam, Science, and the Challenge of History,* New Haven: Yale University Press, 2010.

M. Dols, *Medieval Islamic Medicine: Ibn Riḍwān's Treatise "On the Observation of Bodily Ills in Egypt,"* Berkeley; Los Angeles; London: University of California Press, 1984.

M. Dols, The Origins of the Islamic Hospital: Myth and Reality, *Bulletin of the History of Medicine* 61, 1987, pp. 367–90.

M. Dols, review of *Health and Medicine in the Islamic Tradition: Change and Identity,* by Fazlur Rahman, *History of Science* 74, 1988, p. 420.

F.M. Douglas, Controversy and its Effects in the Biographical Tradition of al-Khaṭīb al-Baghdādī, *Studia Islamica* 46, 1977, pp. 115–31.

H. Ebrahimnajed, The Development of Galenic-Islamic Medicine: Assimilation of Greek Sciences into Islam, *Disquisitions on Past and Present* 13, 2005, pp. 127–40.

N.M. El Cheikh, Gender and Politics in the Harem of al-Muqtadir, in Leslie Brubaker and Julia M.H. Smith, eds., *Gender in the Early Medieval World East and West, 300–900,* Cambridge: Cambridge University Press, 2004, pp. 144–55.

N.M. El Cheikh, Revisiting the Abbasid Harems, *Journal of Middle East Women's Studies* 1/3, 2005, pp. 1–19.

C. Elgood, *A Medical History of Persia and the Eastern Caliphate from the Earliest Times until the year A.D. 1932,* Amsterdam: APA-Philo Press, 1979 (reprint).

G. Endress, The Circle of al-Kindi, in Gerhard Endress and Remke Kruk, eds., *The Ancient Tradition in Christian and Islamic Hellenism,* Leiden: Research School CNWS, 1997, pp. 43–76.

J.M. Fiey, *Chrétien Syriaque sous les Abbassides surtout á Bagdad (749–1258),* Louvain: Secrétariat du Corpus SCO, 1980.

F.B. Flood, *Objects of Translation: Material Culture and Medieval "Hindu-Muslim" Encounter*, Princeton: Princeton University Press, 2009.
H. Gibb, Islamic Biographical Literature, in Bernard Lewis and P.M. Holt, eds., *Historians of the Middle East,* London, New York, Toronto: Oxford University Press, 1962, pp. 54–8.
A. Giladi. *Muslim Midwives.* Cambridge: Cambridge University Press, 2014.
M.H. Green, *Making Women's Medicine Masculine: The Rise of Male Authority in Pre-Modern Gynaecology,* Oxford, New York: Oxford University Press, 2008, pp. 304–9.
R. Gül, Greek Medicine in Persia, *Encyclopaedia Iranica*, XI/4, pp. 342–57; available online at http://www.iranicaonline.org/articles/greece-x (accessed on 26 July 2010).
D. Gutas, *Greek Thought, Arabic Culture: The Graeco-Arabic Translation Movement in Baghdad and Early ʿAbbāsid Society (2nd–4th/ 8th–10th Centuries)*, London and New York: Routledge, 1998.
D. Gutas, *Greek Wisdom Literature in Arabic Translation: A Study of the Graeco-Arabic Gnomologia*, New Haven: American Oriental Society, 1975.
D. Gutas, Paul the Persian on the Classification of the Parts of Aristotle's Philosophy: a Milestone between Alexandria and Baġdâd, *Der Islam* 60/2, 1983, pp. 231–67.
G. Hambly, Becoming Visible: Medieval Islamic Women in Historiography and History, in Gavin Hambly, ed., *Women in the Medieval Islamic World: Power, Patronage and Piety,* St. Martin's Press: New York, 1998, pp. 3–27.
P. Horden, How Medicalised Were Byzantine Hospitals?, in *Hospitals and Healing from Antiquity to the Later Middle Ages,* Variorum Collected Studies Series, Aldershot: Ashgate, 2008.
P. Horden, What's Wrong with Early Medieval Medicine?, *Social History of Medicine* 24/1, 2011, pp. 5–25.
Ibn Abī Uṣaybiʿa, Muwaffaq al-Dīn Abū al-ʿAbbās Aḥmad b. al-Qāsim, *ʿUyūn al-anbāʾ fī ṭabaqāt al-aṭibbāʾ*, ed. August Müller, Frankfurt am Main: Institute for the History of Arabic-Islamic Science at the Johann Wolfgang University, 1995 (rep.)
Ibn al-Jawzī, Abū al-Faraj ʿAbd al-Raḥmān b. ʿAlī, *al-Muntaẓam fī tārīkh al-mulūk wa al-umam*, ed. Muḥammad ʿAbd al-Qādir ʿAṭā, et al., 19 vols. in 17, Beirut: Dār al-Kutub al-ʿIlmiyya, 1992.
Ibn Juljul, Abū Dāwūd Sulaymān b. Ḥassān al-Andalusī, *Kitāb ṭabaqāt al-aṭibbāʾ wa al-ḥukamāʾ*, ed. Fuʾād Sayyid, Cairo: Institut français d'archéologie orientale, 1955.
Ibn al-Nadīm, *Kitāb al-fihrist*, ed. Gustav Flügel, Leipzig: Verlag von F.C.W. Vogel, 1872.
Ibn al-Qifṭī, Jamāl al-Dīn Abū al-Ḥasan ʿAlī b. Yūsuf, *Ikhbār al-ʿulamāʾ bi-akhbār al-ḥukamāʾ*, ed. Julius Lippert, Leipzig: Dieterich'sche Verlagsbuchhandlung, 1903.
al-Jāḥiẓ, Abū ʿUthmān ʿAmr b. Baḥr al-Baṣrī, *Kitāb al-bukhalāʾ*, ed. Gerlof Van Vloten, Leiden: E.J. Brill, 1900.
al-Jāḥiẓ, Abū ʿUthmān ʿAmr b. Baḥr al-Baṣrī, *The Book of Misers = al-Bukhalāʾ*, trans. R.B. Serjeant, Reading: Garnet Publishing, 1997.
J. Jouanna, *Hippocrates,* trans. M.B. DeBevoise, Baltimore and London: John Hopkins University Press, 1999.
S.M. Khan, Ali Ibn Rabban aṭ-Ṭabari, a Ninth Century Arab Physician, on the Āyurveda, *Indian Journal of History of Science*, 25, 1990, pp. 1–14.
G. Leiser, *Medical Education* in Islamic Lands from the Seventh to the Fourteenth Century, *Journal of History of Medicine and Allied Sciences* 38, 1983, pp. 48–75.

M. Marín, *Mujeseres en al-Ándalus*, Madrid: Consejo Superior de Investiaciones Cientícas, 2000.
al-Masʿūdī, *Murūj al-dhahab wa-maʿādin al-jawhar = Les prairies d'or*, ed. and trans. C. Barbier de Meynard et Pavet de Courteille, 9 vols., Paris: Impr. Impériale, 1861–1877.
al-Masʿūdī, *The meadows of gold*, ed. and trans. Paul Lunde and Caroline Stone, Kegan Paul International: London, 1989, pp. 387–392.
M. Meyerhof, ʿAlī at-Tabarī's "Paradise of Wisdom:" One of the Oldest Arabic Compendiums of Medicine, *Isis* 16/1, 1931, pp. 6–54.
M. Meyerhof, New light on Ḥunain Ibn Isḥâq and his period, *Isis* 8, 1926, pp. 685–724.
M. Meyerhof, On the Transmission of Greek and Indian Science to the Arabs, *Islamic Culture* 11, 1937, pp. 17–29.
F. Micheau, Mécènes et médecins à Bagdad au IIIe/IXe siècle les commanditaires des traductions Galien par Ḥunayn ibn Isḥāq, in Danielle Jacquart, ed., *Les voies de la science grecque, étude sur la transmission des textes de l'Antiquité au dix-neuvième siècle*, Geneva: Librairie Droz, 1997, pp. 147–80.
V. Nutton, *Ancient Medicine*, London and New York: Routledge, 2004.
V. Nutton, From Galen to Alexander: Aspects of Medicine and Medical Practice in Late Antiquity, in John Scarborough, ed., *Dumberton Oaks Papers* XXXVIII, Washington: Dumberton Oaks Research Library, 1984, pp. 1–15.
D. Pingree, The Greek Influence on Early Islamic Mathematical Astronomy, *Journal of the American Oriental Society* 93/1, 1973, pp. 32–43.
P. Pormann, The Art of Medicine: Female Patients and Practitioners in Medieval Islam, *The Lancet* 373 (9 May 2009), pp. 1598–9.
P. Pormann, Islamic Hospitals in the Time of al-Muqtadir, in J. Nawas, ed., *Abbasid Studies II: Occasional Papers of the School of 'Abbasid Studies, Leuven, 28 June–1 July 2004*, Orientalia Lovaniensia Analecta 177, Leuven, Dudley: Peeters, 2010, pp. 337–82.
P. Pormann, Medical Methodology and Hospital Practice: The Case of Tenth-century Baghdad, in Peter Adamson, ed., *In the Age of al-Farabi: Arabic Philosophy in the 4th/10th Century*, Warburg Institute Colloquia 12, London: Warburg Institute, 2008, pp. 95–118.
P. Pormann, The Physician and the Other: Images of the Charlatan in Medieval Islam, *Bulletin of the History of Medicine* 79/2, 2005, pp. 189–227.
P. Pormann and E. Savage-Smith, *Medieval Islamic Medicine*, Washington: Georgetown University Press, 2007.
W. al-Qadi, Biographical Dictionaries as the Scholars' Alternative History of the Muslim Community, in Gerhard Endress, ed., *Organizing Knowledge: Encyclopaedic Activities in the Pre-Eighteenth Century Islamic World*, Leiden: Brill, 2006, pp. 23–75.
W. al-Qadi, Biographical Dictionaries: Inner Structure and Cultural Significance, in George N. Atiyeh, ed., *The Book in the Islamic World: The Written Word and Communication in the Middle East*, New York: State University of New York Press, 1995, pp. 93–122.
F. Rosenthal, *A History of Muslim Historiography*, Leiden: Brill, 1968.
F. Rosenthal, The Physician in Medieval Muslim Society, *Bulletin of the History of Medicine* 52, 1978, pp. 476–91.
J.R. Russell, Christianity in Pre-Islamic Persia: Literary Sources, *Encyclopaedia Iranica*, V/5, pp. 523–8; available online at http://www.iranicaonline.org/articles/christianity-i (accessed on 26 July 2010).

E. Savage-Smith, Ṭibb, in Peri Bearman, Th. Bianquis, Clifford Edmund Bosworth, E. van Donzel, and W.P. Heinrichs, eds., *Encyclopaedia of Islam,* 2nd ed., Leiden: Brill, 2010, Brill Online, accessed 29 September 2010, http://www.brillonline.nl/subscriber/entry?entry=islam_COM-1216.

M. Shatzmiller, Aspects of Women's Participation in the Economic Life of Later Medieval Islam: Occupations and Mentalities, *Arabica* 35/1, 1988, pp. 36–58.

M. Shefer-Mossensohn, A Sick Sultana in the Ottoman Imperial Palace: Male Doctors, Female Healers and Female Patients in the Early Modern Period, *Hawwa* 9/3, 2011, pp. 281–312.

D. Sourdel, Bu͟kh͟tīs͟h͟ū', in Peri Bearman, Th. Bianquis, Clifford Edmund Bosworth, E. van Donzel, and W.P. Heinrichs, eds., *Encyclopaedia of Islam,* 2nd ed., Leiden: Brill, 2010, Brill Online, accessed 06 December 2010, http://www.brillonline.nl/subscriber/entry?entry=islam_SIM-1514.

G. Strohmaier, Ḥunayn b. Isḥāḳ al-'Ibādī, in Peri Bearman, Th. Bianquis, Clifford Edmund Bosworth, E. van Donzel, and W.P. Heinrichs, eds., *Encyclopaedia of Islam,* 2nd ed., Leiden: Brill, 2010, Brill Online, accessed 14 December 2010, http://www.brillonline.nl/subscriber/entry?entry=islam_COM-0300.

al-Tha'ālibī, 'Abd al-Malik b. Muḥammad, *Ghurar akhbār mulūk al-Furs = Histoire des rois des Perses,* ed. and trans. H. Zotenberg, Paris: Impr. nationale, 1900 [Tehran: M.H. Asadi, 1963], pp. 531–32.

al-Ṭabarī, Abū Ja'far Muḥammad b. Jārīr, *Ta'rīkh al-rusul wa al-mulūk*, ed. Michael Johan De Goeje, 15 vols., Lugduni-Batavorum: E..J. Brill, 1964–1965.

al-Ṭabarī, Abū Ja'far Muḥammad b. Jārīr, *The History of al-Ṭabarī*, vol. 29 (*Al-Manṣūr and al-Mahdī*); trans. and annot. Hugh Kennedy, Albany: SUNY Press, 1990.

al-Ṭabarī, 'Alī b. Rabbān, *Firdaws al-Ḥikmah, ed.* M. Zubayr al-Sidiqi, Berlin: Maṭbat Āftāb, 1928.

L.M.V. Totelin, *Hippocratic Recipes: Oral and Written Transmission of Pharmacological Knowledge in fifth and fourth Century Greece*, Leiden: Brill, 2008.

P. Unschuld, The Physician in Imperial China: Medical Ethics and Malpractice Legislation, in Teizo Ogawa, ed., *History of Professionalization of Medicine*, Osaka: Division of Medical History, The Taniguchi Foundation, 1987, pp. 90–115.

J.-C. Vadet, Ibn Māsawayh, Abū Zakariyyā Yuḥannā, in Peri Bearman, Th. Bianquis, Clifford Edmund Bosworth, E. van Donzel, and W.P. Heinrichs, eds., *Encyclopaedia of Islam,* 2nd ed., Leiden: Brill, 2010, Brill Online, accessed 08 January 2011, http://www.brillonline.nl/subscriber/entry?entry=islam_SIM-3289.

A. Van Arsdall, Medical Training in Anglo-Saxon England: an Evaluation of the Evidence, in Patrizia Lendinara, Loredana Lazzari, and M.A. D'Aronco, eds., *Form and Content of Instruction in Anglo-Saxon England,* Turnhout: Brepols, 2007, pp. 415–43.

B. Wilson, *The Bakhtishu': Their Political and Social Role under the 'Abbasid Caliphs (A.D. 750–1100)*, PhD diss., New York University, 1974.

M.J.L. Young, Arabic Biographical Writing, in M.J.L. Young, J.D. Lathman, and R.B. Serjeant, eds., *Religion, Learning and Science in the Abbasid Period*, Cambridge: Cambridge University Press, 1990, pp. 168–87.

Charles Burnett and David Juste

A New Catalogue of Medieval Translations into Latin of Texts on Astronomy and Astrology

This article summarizes some preliminary results from a catalogue of medieval translations of Latin texts on astronomy and astrology, which we have been engaged in since 2009. For over fifty years Francis J. Carmody's *Arabic Astronomical and Astrological Sciences in Latin Translation*[1] has served as the first point of call for identifying Latin translations of Arabic astrological and astronomical texts and the manuscripts in which they occur. There is an obvious need not so much to update this book as to replace it entirely, taking into account the new research and discoveries that have been made in the meantime – the fuller and more informed catalogues of manuscript collections and personal inspection on the original or on reproductions – and applying much more rigorous methods of scholarship.

The new, critical catalogue is intended to be published both online and in printed form. It includes all known medieval Latin translations of astronomical and astrological texts – not only the translations from Arabic, but also those from Greek and Hebrew, which Carmody neglected – , and, as far as possible, lists all manuscripts and early printed editions. The texts surveyed were translated primarily in the period between 1100 and 1400, though the manuscripts are often much more recent. The printed catalogue is arranged in alphabetical order and includes a brief biography and bibliography, followed by the works of that author that were translated into Latin. For each work the following information is provided: (1) the title, incipit and explicit, quoted from representative manuscripts; (2) a brief description of the nature and contents of the work; (3) the identification of the Arabic, Greek or Hebrew original, as well as the date, place and circumstances of the translation; (4) a bibliography; (5) the available editions, early modern and modern; and (6) a full list of manuscripts. At the end of the catalogue anonymous texts will be provided, as well as Latin texts for which a Greek, Arabic, or Hebrew original is suspected, but not demonstrable. Indices of manuscripts, of Latin translators, and of incipits will appear in the printed catalogue; and an online version will of course make available several other ways of accessing and searching the material.

---

1 Published in Berkeley, 1955, by the University of California Press.

This catalogue will draw on several publications which have appeared since Carmody's work, including Lynn Thorndike and Pearl Kibre's *A Catalogue of Incipits of Mediaeval Scientific Writings in Latin*;[2] Fuat Sezgin's *Geschichte des arabischen Schrifttums*, vols. 6 (*Astronomie*) and 7 (*Astrologie – Meteorologie und Verwandtes*);[3] and, above all, David Juste's own catalogue of Latin astrological manuscripts, which is in the process of being written and published.[4] Online resources such as *Jordanus, an International Catalogue of Mediaeval Scientific Manuscripts*[5] and the German *Manuscripta Mediaevalia*[6] have also been used. As far as possible, however, entries will be based on direct consultation of the manuscripts involved – a task greatly facilitated by the microfilm archive in the Institute for the History of Science at the University of Munich, where the project was begun.[7]

Although we now have a document of over 135,000 words, the catalogue is far from being finished;[8] but it is time to take stock of what has been achieved and what we can deduce from its contents. We will concentrate on the nature of the three sources of Latin translations we are dealing with: Greek, Arabic, and Hebrew.

---

2 2nd ed. published in London, 1963, by the Mediaeval Academy of America.

3 Both volumes in Leiden, 1978 and 1979, by E.J. Brill.

4 *Catalogus Codicum Astrologorum Latinorum*, Paris, CNRS Editions, vol. I (2011) and vol. II (2015).

5 Now hosted by the project Ptolemaeus Arabus et Latinus at the Bavarian Academy of Science and Humanities (Munich), at the address jordanus@ptolemaeus.badw.de.

6 Supported by the Deutsche Forschungsgemeinschaft, http://www.manuscripta-mediaevalia.de.

7 The microfilm archive has now been transferred to the Monumenta Germaniae Historica (Munich). Also proving very useful for us at the Institute were: a photocopy of a copy of Carmody's text that had been heavily annotated – with the addition of many new manuscripts – by Francis Benjamin; other copies of Carmody annotated by Paul Kunitzsch and Heinrich Hermelink; a copy of Ernst Zinner's *Verzeichnis der astronomischen Handschriften*, annotated by Zinner himself; and a handwritten catalogue (in the form of fiches) of medieval Latin mathematical manuscripts, compiled, but never published, by Axel Bjørnbo. This is not to mention the helpfulness of the personnel in the Institute – above all professor Menso Folkerts, but also Paul Kunitzsch, Richard Lorch and Benno van Dalen, and the Arabist François Charette, who was able to join us for a fortnight in 2009.

8 Note that all the figures given in this article are valid as of July 2015, but are likely to change as the catalogue develops.

We have altogether some sixty authors: eight of these are Greek, whose works are sometimes translated directly from Greek, at other times via Arabic;[9] six write in Hebrew; and, aside from one Zoroastrian (whose work would originally have been written in Middle Persian), the rest of the original authors wrote in Arabic – though among them were Jews and Christians, as well as Muslims. The number of manuscripts and early printed editions of any one of the texts we have catalogued ranges from one (or even zero, if one counts translations no longer extant), to over two hundred. In assessing the significance of this, one has of course to take into account the fact that some manuscript collections are better described than others,[10] and some libraries from the Middle Ages have been destroyed.[11] Moreover, some collections have remained intact, while others have been dispersed.[12] Nevertheless, the numbers and the distribution of manuscripts of texts can tell us something.

Let us take first the Greek sources, whose authorities are, without exception, the ancient Greeks. Some texts, e. g. the *Excerpts from the Secrets of Albumasar* (Shādhān b. Baḥr's anecdotes about his master Abū Maʿshar Jaʿfar b. Muḥammad b. ʿUmar al-Balkhī d. ca. 886 AD), and a major text on the revolution of nativities by the Abū Maʿshar himself, have been translated into Latin via a Greek intermediary text of ca. 1000 AD; but the work of no late Hellenistic or Byzantine astronomer or astrologer (although such scholars existed) was translated into Latin in the Middle Ages. The Greek authorities, rather, range from Autolycus of Pitane (fourth century BC) to Ptolemy (second century AD).

A Greek curriculum in astronomy, which began with Euclid's *Elements* and ended with Ptolemy's *Almagest*, was established in Alexandria. It is referred to and summarised by the early fourth-century mathematician Pappus of Alexandria, and included several shorter texts on spherical geometry, and astronomical observation and calculation, which Pappus called the "lesser astronomy."[13] In-

9 We do not count Dorotheus, Azareus, Thetel, and texts attributed to Aristotle, Hermes, Hippocrates and Salomon, most of which are unlikely to have any Greek source.

10 For example, the scientific manuscripts of the Jagellonian Library in Krakow and the Palatine Collection of the Vatican have been described in detail by Grażyna Rosińska (*Scientific Writings and Astronomical Tables in Cracow*) and Ludwig Schuba (*Die Quadriviums-Handschriften der Codices Palatini Latini in der Vatikanischen Bibliothek*).

11 This includes the libraries of Chartres and Metz.

12 For example, the manuscripts of the Oxford Colleges and the Bodleian Library tend to be those that were copied and/or used in the same colleges and the University of Oxford in the Middle Ages, whilst those in the British Library, the Vatican Library, and the Staatsbibliothek at Berlin have come from many different and widely scattered sources.

13 See Pingree, review of *Hypsikles*, pp. 15 – 16. Otto Neugebauer (*History of Ancient Mathematical Astronomy*, pp. 768 – 9) expresses reservations concerning the term.

cluded were Theodosius's *Spherics* and *On Days and Nights*, Autolycus's *Moving Sphere*, Aristarchus's *On the Size and Distances of the Sun and Moon*, and Euclid's *Optics and Phaenomena*. A slightly larger programme (with the addition of Theodosius's *On Inhabitable Places*, Autolycus's *On Rising and Setting*, and Hypsicles's *On Ascensions*) was evidently known to Ḥunayn b. Isḥāq, the most prolific translator of Greek (and Syriac) scientific and medical writings into Arabic in the ninth century; and his translations of these texts are extant in Arabic. The same programme was known to Gerard of Cremona, who, in Toledo, in the second half of the twelfth century, played an equivalent role to that of Ḥunayn by transmitting Arabic scientific and medical writings into Latin. In the largest collection of his mathematical translations is a note that "according to what is found in the writings of Johannitius (i.e., Ḥunayn b. Isḥāq) this is the order of works which (should be read) after Euclid;" the full list of the "lesser astronomy" follows.[14] One can see here, then, an opportunity for the ancient Greek astronomical curriculum, of considerable mathematical sophistication, to pass to the Latins through the medium of Arabic; and, perhaps, it was supplemented by direct translations from Greek, as happened in the case of Aristotelian natural philosophy and some other areas of geometry and medicine. But what do the manuscripts tell us concerning what actually happened?

Only in the case of Euclid's *Optics* do we have translations from both Greek and Arabic.[15] Of the other eight texts we have translations from Arabic only (in the Middle Ages), including four made by Gerard of Cremona: Theodosius's *Spherics* and *On Habitable Places*, Autolycus's *On the Moving Sphere*, and Hypsicles's *On Ascensions*; but of these, respectively, only 20, 11, four, and five manuscripts have survived. *On Spherical Figures* by Menelaus of Alexandria, a related text on spherical geometry also translated by Gerard of Cremona, has survived in 20 manuscripts; and another, anonymous, version of Theodosius's *Spherics* is extant in 14 manuscripts, because it was attached to the most popular version of Euclid's *Elements*.

We might add those texts of Arabic mathematicians that were developments of, or were inspired by, the ancient Greek texts: such as Thābit b. Qurra's *On the Sector Figure*, and Aḥmad b. Yūsuf's *On Similar Arcs* (both of which, again, were

14 In manuscript Paris, Bibliothèque nationale de France lat. 9335, f. 28 verso. This passage has most recently been edited in Burnett, Coherence, 274–5.

15 See Takahashi, Mori, and Kikuchihara, Paraphrased Latin Version of Euclid's *Optica*, pp. 127–92, concerning a text that brings together the Greek-Latin and Arabic-Latin versions of the *Optics*. The *Optics* formed part of a group of Greek texts studied by the anonymous translator of the *Almagest* from Greek into Latin, in Sicily, ca. 1160: Euclid's *Data*, *Optica*, and *Catoptrica*, and Proclus's *Elementatio physica* (Haskins, *Studies*, p. 191).

translated by Gerard). The latter is extant in seven manuscripts, the former in four (to which one can add five more from other versions); but we are dealing with a very small number of copies of texts made over a period of some 350 years. We cannot, therefore, say that this Hellenistic and Arabic astronomical programme caught on in the West.

The *Almagest* itself was a great draw for the translators, and three twelfth-century translations survive, one from Greek and two from Arabic. But the first one, made by ʿAbd al-Masīḥ of Winchester, probably in Antioch, survives in one manuscript alone, containing only the first four books;[16] a second, made directly from the Greek, fared little better in that it is found complete in two manuscripts, and incomplete in five others.[17] Even Gerard's translation, of which two versions exist (one original and one revised), does not exceed 58 copies.

One reason for this is that a different programme in astronomy developed in Western schools, dependent on a continuous and developing tradition in the Islamic world, especially in al-Andalus – Islamic Spain – , with which the Latin translators had the closest contact. In al-Andalus we see a succession of astronomers, from Maslama al-Majrīṭī (turn of the 10th to 11th centuries), who revised and improved astronomical tables and constructed astrolabes; to his pupil Ibn al-Ṣaffār (d. 1035), who wrote an influential book on how to use an astrolabe; to Ibn Muʿādh in Jaén and al-Zarqālī (al-Zarqālluh) in Toledo, who were both emending astronomical tables; to Jābir b. Aflaḥ of Seville (early 12th century), who provided an up-to-date and easier-to-understand version of Ptolemy's *Almagest*; to al-Biṭrūjī (late twelfth century), who criticized the Ptolemaic model on the grounds of its incompatibility with Aristotelian physics. Works of all these authors were translated into Latin – usually soon after they were composed in Arabic – from the earliest texts on the astrolabe in Latin (in late tenth century Catalonia), which used Maslama's star-table,[18] to Michael Scot's early thirteenth-century translation of the *Book on the Configuration (of the World)* by al-Biṭrūjī (Alpetragius).

Because works of later generations of Arabic astronomers superseded the works of earlier ones, their translations into Latin in turn displaced the translations of their predecessors' works. Thus many astronomical works survive in Latin only in one or two manuscripts, and those manuscripts tend to belong

---

16 In manuscript Dresden, Sächsische Landesbibliothek – Staats- und Universitätsbibliothek, Db. 87, ff. 1 recto – 71 recto. See Burnett, ʿAbd al-Masīḥ of Winchester, pp. 159 – 69. An edition and translation of this text is being prepared by Dirk Grupe.

17 The "Sicilian" *Almagest*, discussed in Haskins, *Studies*, pp. 157 – 64 and 191 – 3.

18 See Samsó, Maslama al-Majrīṭī, pp. 506 – 22.

to a time close to the period of translation. A prime example of revision and development is provided by astronomical tables, from which the movement of Sun, Moon, and planets could be calculated for any time and any place. The Tables of al-Khwārizmī, translated in the early twelfth century, were replaced by the Tables of Toledo (at least two successive versions introduced from the Arabic), and eventually by the Alfonsine Tables, which were largely an indigenous product of the Castilian astronomers of the late thirteenth century.[19] But in between these there were several other Arabic tables introduced into Latin with their instructions (or 'canons'), which tend to be witnessed by only a few manuscripts close in date to the time of translation. Al-Khwārizmī's canons and tables are found in three versions, in ten manuscripts altogether, all written in the twelfth and thirteenth centuries (except one from s. XIV – XV). The *Canons of the Tables of Humeniz (Arzachel?)*, translated in the thirteenth century by John of Pavia, are extant in eight manuscripts, of which six are from the same thirteenth century.

This lack of diffusion can be seen in other Arabic astronomical works, too: of Ibn al-Haytham's *On the Configuration of the World*, bringing together Ptolemaic astronomy and cosmology, three translations are known from the twelfth and thirteenth century, but all are extant in only one or two manuscripts; and the twelfth-century translation (*Liber Mamonis)* is nearly contemporary with its sole manuscript.[20] Ibn al-Muthannā's commentary on the Tables of al-Khwārizmī may have been translated within fifty years of its composition, but the translation survives in only three manuscripts, two of which could be contemporary with the translator, Hugo of Santalla.

The displacement of older Arabic translations by newer translations is one phenomenon; however, because of developments in Latin astronomy, the invention of new and more sophisticated astronomical instruments, and the results of direct observation of the heavens, the translations of Arabic astronomical works in turn became displaced by indigenous Latin texts. This can be seen especially in the teaching in the universities, where, in the mid-thirteenth century, texts by John of Sacrobosco and Petrus Philomena, and a *Theory of the Planets* (*Theorica planetarum*) of disputed authorship, while all being directly based on Arabic material, became the standard texts on time-reckoning, and on the movements of the Sun, Moon, and planets.[21] In the field of instrument making, a text on the

19 Chabás and Goldstein, *The Alfonsine Tables of Toledo.*

20 *Incipit liber Mamonis in astronomia, a Stephano philosopho translatus*, MS Cambrai, Médiathèque municipale de Cambrai, no. 930. An edition and translation of this text will soon be published by Dirk Grupe.

21 See Pedersen, The Corpus Astronomicum, pp. 57 – 96.

uses of the astrolabe, purporting to be by an Arab, Māshā'allāh, was cobbled together from preexisting Latin translations, and became the standard text;[22] while the sophisticated quadrant, also of Arabic origin, was the subject of a text by Robert the Englishman ("the old quadrant"), until this was replaced by a new quadrant invented and described by the Hebrew scholar Prophatius. These university texts existed in hundreds of manuscript copies; the development of this corpus was largely due to the composition of more and more advanced commentaries on these basic texts. But, as one can see, this is completely different from the Greek curriculum of astronomy which, as the manuscripts show, failed to catch on in medieval European universities.

The substitution of indigenous Latin texts on astronomy for Latin translations of Arabic texts is evidenced by the dearth of astronomical manuscripts in our catalogue, and the rapid decline of the copying of Latin translations of Arabic texts after the time of their original translation. There are two notable exceptions to this. The first is al-Farghānī's *Thirty Chapters* (*On the Science of the Stars*). This summarised Ptolemaic astronomy but added new information (some of it relevant to astrology), including material on the twenty-eight lunar mansions, the elongations of planets and fixed stars from the earth, the occultation of the planets by the Sun, and their being "under the rays of the Sun." This text had already functioned as a teaching text in Arabic, and had received commentaries, just as Sacrobosco's *On the Sphere* received commentaries in Latin; it must have become known to Latin translators as an introduction to (if not a substitute for) Ptolemy's *Almagest*. We have two twelfth-century Latin translations, by John of Seville and Gerard of Cremona, respectively, yielding 117 manuscripts altogether.

The other exception is a group of short texts attributed to Thābit b. Qurra. These include *On those things that one needs to know before reading the Almagest* (*De hiis que indigent expositione antequam legatur Almagesti*; in 67 manuscripts) and *On the Correct Envisaging of the Sphere and its Different Circles* (*De recta imaginatione spere et circulorum eius diversorum*; in 71 manuscripts). A third text, *On the Sizes of the Stars and Planets and their Ratio to the Earth* (*De quantitate stellarum et planetarum et proportione terre*; in 52 manuscripts) included in this group is clearly not a version of a text by Thābit, and may not even be a translation, while a fourth text, again, is not by Thābit–*On the Motion of the Eighth Sphere* (*De motu octave spere*; in 110 manuscripts). This fourth text is devoted to a description of an element of celestial movement ('trepidation') that is

22 Kunitzsch, *Glossar der arabischen Fachausdrücke*, pp. 499–500. See now Pseudo-Māshā'allāh, On the Astrolabe, ed. Ron B. Thomson, Version 1.2 (Toronto, 2015), available online.

not Ptolemaic, but coincides with the theory put forward in the *Theory of the Planets*, and was therefore the norm in the Middle Ages – until it was severely criticized by the "reviver of Ptolemy," Johannes Regiomontanus, in his famous *Disputations against the Cremonese's Ravings* (*Disputationes contra Cremonensia deliramenta*), of ca. 1465.

A rather different picture emerges when one looks at the astrological works and authorities. Several ancient Greek authorities are evoked (but not, as we have said, late Hellenistic or Byzantine ones) but the only ancient texts translated were those of Ptolemy. Six translations of the *Tetrabiblos* were made in the twelfth and thirteenth centuries. Four of those survive in one to four manuscripts only, but two were more successful; namely, Plato of Tivoli's translation from 1138 (39 manuscripts), and Aegidius of Tebaldis's translation, made at the court of Alfonso X between 1271 and 1275 (45 manuscripts). The latter includes the extensive eleventh-century commentary by the Cairene doctor ʿAlī b. Riḍwān. The *Centiloquium*, which may go back to a Greek source, was also known through a commentary made by Aḥmad b. Yūsuf in Cairo, in the mid-tenth century. The text survives in five twelfth-century translations (165 manuscripts altogether), the most popular of which was due, once again, to Plato of Tivoli, and was made in 1136 (over 100 manuscripts).

The vast majority of astrological works in our catalogue are by Arabic authors who lived in the eighth and ninth centuries. And these texts, as one would expect for these dates, come from the central area of the Islamic realm, and not from al-Andalus (as do so many of the later astronomical texts). Thus we have, from the eighth and early ninth century, astrologers from Persia, or with a Persian ancestry: al-Andarzaghar, al-Faḍl b. Sahl, Māshāʾallāh (together with his student, Abū ʿAlī al-Khayyāṭ), ʿUmar b. al-Farrukhān , Abū Bakr, and Sahl b. Bishr. From the ninth century we have writers on astrology who were contemporary with Ḥunayn b. Isḥāq, and involved in the transmission of Greek scientific and philosophical learning in Baghdad: al-Kindī, Abū Maʿshar and his pupil Shādhān b. Baḥr. And from the tenth century we have al-ʿImrānī of Mosul and his pupil al-Qabīṣī, whose patron was Sayf al-Dawla, the Emir of Aleppo from 945–67, and Aḥmad b. Yūsuf (d. ca. 941), in the Tulinid court in Cairo. The only later writers are a certain Isrāʾīlī who addressed a set of astrological aphorisms to the Fatimid caliph, Abū ʿAlī al-Manṣūr (r. 996–1021) in Cairo, ʿAlī b. Riḍwān (d. 1068) the chief physician of Egypt, who served the Fatimid caliph al-Mustanṣir, and ʿAlī b. Abī al-Rijāl (d. after 1040), who served the Zirids in Kairouan. As we have seen, the later authors differed from those of the eighth and ninth centuries in that they wrote commentaries (on Pseudo-Ptolemy's *Centiloquium*, and on Ptolemy's *Quadripartitum*); but they also wrote summaries (Isrāʾīlī), or, in the case of ʿAlī b. Abī al-Rijāl, a vast, all-inclusive compendium, rather than textbooks on the individual branches of astrology. But even these later authors lived some two centuries before the Latin translations were made, and none of them were local to al-Andalus.

The reason for this must be the conservatism of astrology. Its doctrines do not change and develop in the course of the Middle Ages and owe their validity to their purported hoary tradition. And this situation does not change when the texts are translated into Latin. The Latin translations, too, remain unchanged over generations and centuries. The pattern of diffusion, in fact, shows a contrasting picture to that of most astronomical texts: the number of copies does not diminish, but rather grows exponentially from the time of translation (twelfth and thirteenth centuries) until it reaches a peak in the fifteenth century.[23] In fact, there are some texts that can no longer be found in manuscripts close in date to the time of their composition, but only survive in fifteenth- or sixteenth-century copies or prints, such as the introduction to the *Capitula Almansoris* made by Plato of Tivoli in Barcelona in 1136, and the translations of Salio of Padua made in Toledo in the 1210s and 1220s. When astrological texts started to be printed from the 1470s onwards, most of the texts printed were Latin translations made in the twelfth and thirteenth centuries, of Arabic works written in the eighth or ninth centuries. Thus, astrological texts were not superseded and displaced either in the Arabic or in the Latin traditions.

Medieval and Renaissance scholars left commentaries on some astrological texts. The best examples are for Ptolemy's *Quadripartitum*, Pseudo-Ptolemy's *Centiloquium* and, especially, al-Qabīṣī's *Introduction to Astrology*. That the latter received several commentaries (we know of at least eight) is not surprising since, as Olaf Pedersen pointed out, it was included in the *Corpus astronomicum* since at least the late fourteenth century, and became the subject of university study.[24] One commentary – that of John of Stendhal – is described specifically as "for the students of Erfurt," in 1359. The most commonly occurring commentary – in over 30 manuscripts and eight Renaissance editions[25] – was written by John of Saxony in 1331, who also wrote the most popular set of instructions for the Alfonsine Tables (also on the university curriculum).

Some texts proved more popular than others. Hugo of Santalla's translation of the *Book of Aristotle Containing the Sum of the 255 Volumes of the Indians* – a

23 See Juste, The Impact of Arabic Sources on European Astrology, p. 183.

24 Pedersen, The Corpus Astronomicum, pp. 80 – 1. See also Burnett, Al-Qabisi's *Introduction to Astrology*, pp. 43 – 69; Burnett, Teaching the Science of the Stars in Prague.

25 The manuscripts are listed in *al-Qabisi (Alcabitius): The Introduction to Astrology*, pp. 509 – 10, the editions are on pp. 192 – 4.

translation of an eighth-century Arabic work by Māshā'allāh which, in turn, is apparently based largely on Middle Persian material – survives in a manuscript contemporary with the translator, and one later copy of the same manuscript. Other examples are two versions of Māshā'allāh's *On Nativities*, which both occur in two manuscripts only. But others were extremely popular, as the following list will show:

- al-Qabīṣī, *Introduction to Astrology:* 211 MSS
- *Astronomia Ypocratis* (on astrological medicine): 138 MSS (3 translations)
- Abū Ma'shar, *Flowers:* 124 MSS
- Sahl b. Bishr, *Book on the Judgements of the Stars:* 122 MSS
- Māshā'allāh, *Book on the Significance of Eclipses:* 117 MSS
- 'Alī b. Abī al-Rijāl, *On the Judgements of the Stars:* 99 MSS
- Ptolemy's *Quadripartitum:* 94 MSS (6 medieval translations)
- Gergis, *On the Significance of the Seven Planets in the Houses:* 65 MSS
- Abū Ma'shar, *Great Introduction:* 59 MSS (2 translations)
- Māshā'allāh, *Book of Reception:* 59 MSS

These texts, sometimes within a single text, sometimes separately, covered the several well-defined parts of astrology: introduction (providing definitions of terms), natal astrology (nativities and their revolutions), interrogations, elections, general astrology (often including weather forecasting as a separate category), and medical astrology. But particularly popular were texts of astrological "aphorisms:"[26]

- *Centiloquium Hermetis:* 86 MSS
- Sahl b. Bishr, *50 precepts* (second part of his *Book on the Judgements of the Stars*): 84 MSS
- *Capitula Almansoris* (150 aphorisms): 79 MSS
- *Centiloquium Bethen:* 43 MSS
- And above all, the *Centiloquium Ptolomei* (of which 5 translations were made in the twelfth and thirteenth centuries): 165 MSS altogether.

---

26 Typical of aphorisms are the following from the beginning of the *Centiloquium Bethen:* "(2) Know that when planets are retrograde, it is like a man who is sick, dumbstruck and worried; (3) a cadent star is like a dead man, without any movement; (4) a burnt star is like a captive, imprisoned, punished man, without power; (5) stationary before retrogression is like a healthy man from whom health is receding, but a little of it remains... (7) Besieged between two malefics is like a frightened man caught between two enemies." (translated from the text in *Firmicus* et. Basel, Iohannes Hervagius, 1533, second part, p. 89).

The last four sets of aphorisms were included in Hervagius's 1533 edition of the major astrological works of the time (with Firmicus Maternus and Manilius).

One may briefly mention the third source of Latin translations: Hebrew. As we have seen, the translations from Greek are of ancient authors, while those from Arabic tend to be later authors in astronomy, but from the first centuries of Islamic culture for astrology. In the case of the authors of Hebrew texts, these Jewish scholars were already in the West, living and working in the same contexts as the Latin scholars. In fact, we often find that the Latin translation is made in collaboration with the Jewish scholar. This is the case with several astronomical and astrological works of Abraham b. ʿEzra, the Jewish polymath who emigrated from Islamic Tudela to Pisa, and from there travelled to Jewish communities in the South of France, Normandy and, eventually, London, dying in ca. 1160.[27] It is also the case with astronomical tables, drawn up first in Hebrew, and then translated into Latin, by Immanuel ben Jacob Bonfils (Tables of Tarascon, 1365), and by Jacob ben David ben Yom Tov (Bonjorn), who drew up the astronomical tables of Perpignan in 1361. Another Yom Tov, either father or son of this previous scholar, wrote a text on astrology for doctors. The Latin version, *Tractatus Davidi Iudei*, cannot much postdate the Hebrew. And the Latin translation of the *New Quadrant* of Prophatius Judaeus (Jacob ben Machir ibn Tibbon) was translated by Armengaud of Prophatius "secundum vocem" (i.e., on his dictation), two years after the original Hebrew version was written (in 1288), and very soon became part of the *Corpus astronomicum* teaching in the universities, as we have already indicated.

Aside from a large number of new manuscripts – and more details about the works involved, the original authors, and the translators – what new information has emerged as a result of compiling this catalogue? We could point out many texts which have not been recognised before. One of these is *On the Revolutions of the Years of the World* by al-Kindī, translated by Henry Bate of Malines from Hebrew into Latin in 1278, which occurs in a single Vatican manuscript.[28] This text has hitherto been known only in an Arabic fragment (two chapters on rains and on winds), and in a brief quotation in Abraham b. ʿEzra's *Book of*

27 A similar situation is found for the two Hebrew scholars, both of the early twelfth century: Petrus Alfonsi, brought up in Islamic Huesca, and introducing Islamic astronomy to Latin scholars in France and England (this time as a converso); and Bar Ḥiyya, also from Huesca, and collaborating with Plato of Tivoli in Barcelona.

28 MS Vatican, *Biblioteca Apostolica Vaticana*, Pal. lat. 1407, s. XIV–XV, f. 58 recto–62 recto.

the World.[29] For al-Kindī, astrology was very much part of the total world picture of largely Peripatetic physics and cosmology which he described in a large number of treatises and longer textbooks, and any newly discovered work of his is bound to be worth studying. Another new text is an abbreviation of work on interrogations by Aomar (i.e., ʿUmar b. al-Farrukhān), printed by Luca Gaurico, together with Aomar's *On Nativities*, in two editions in 1503 and 1525. The colophon reveals the translator and his Jewish assistant, and the time of the translation:

> Aomar the Tiberiad is completed, with God's praises and help, translated by Master Salomon from Arabic into Latin, with the help of Ibn Abaumet, a Jew, and a good man, in the course of the year from the incarnation of Christ, 1217, the fifth Indiction, the beginning of the third day of August, in the year 613 of the Arabs, the last day of the month called Rabiʿ Thani.

Because of the coincidence of dates and places, and Jewish collaboration, it is evident that 'Salomon' is a corruption of the name 'Salio' (Salio of Padua), from whom we have three other translations of works on astrology and geomancy.[30] A third new text is what appears to be the original Latin version of a set of astrological judgements attributed to Gergis, passages of which have been excerpted for inclusion in the *Book of the Nine Judges.*[31]

These, then, are the preliminary gleanings from this new catalogue of astronomical and astrological texts in Latin translation. As more material is added to the catalogue, more information will be able to be derived from it, and our knowledge of the transmission of science and its transformation of European learning will continue to be enriched.

Bibliography

ʿAbd al-Masīḥ of Winchester, translation of Ptolemy's *Almagest*, Books I – IV, MS Dresden, *Sächsische Landesbibliothek*, Db. 87, ff. 1 recto – 71 recto.

G. Bos and C. Burnett, *Scientific Weather Forecasting in the Middle Ages: The Writings of Al-Kindi*, London: Kegan Paul International, 2000.

29 See G. Bos and C. Burnett, *Scientific Weather Forecasting in the Middle Ages*, pp. 46 – 7 and 422 – 33.

30 See Burnett, *De meliore homine*, pp. 295 – 326.

31 The text is entitled the "Liber Amitegni" and survives in a single Oxford manuscript (Laud. Misc. 594, s. XIV, ff. 142 recto–144 recto). See Burnett, Amitegni: A Newly Discovered Text on Astrological Judgements.

C. Burnett, ʿAbd al-Masīḥ of Winchester, in Lodi Nauta and Arie Vanderjagt, eds., *Between Demonstration and Imagination: Essays on the History of Science and Philosophy Presented to John D. North*, Leiden: Brill, 1999, pp. 159–169.

C. Burnett, Al-Qabisi's *Introduction to Astrology:* From Courtly Entertainment to University Textbook, in Resianne Fontaine, Ruth Glasner, Reimund Leicht, and Giuseppe Veltri, eds., *Studies in the History of Culture and Science: A Tribute to Gad Freudenthal*, Ledin: Brill, 2011, pp. 43–69.

C. Burnett, Amitegni: A Newly Discovered Text on Astrological Judgements, *Bruniana & Campanelliana* 21/2, 2015. pp. 653–662.

C. Burnett, The coherence of the Arabic-Latin Translation Program in Toledo in the Twelfth Century, *Science in Context* 41, 2001, pp. 249–88 [reprinted as article VII in *idem, Arabic into Latin in the Middle Ages: The Translators and Their Intellectual and Social Context* (Variorum Collected Studies Series, 939), Farnham, Eng., and Burlington, Vt.: Ashgate, 2009].

C. Burnett, *De meliore homine*, 'Umar ibn al-Farrukhan al-Tabari on Interrogations. A Fourth Translation by Salio of Padua?, in Antonella Sannino, Pasquale Arfé, and Irene Caiazzo, eds., *Adorare caelestia, gubernare terrena, Atti del Colloquio Internazionale in onore di Paolo Lucentini (Napoli, 6–7 Novembre 2007)*, Turnhout: Brepols, 2011, pp. 295–326.

C. Burnett, Teaching the Science of the Stars in Prague University in the Early Fifteenth Century: Master Johannes Borotin, *Aither. Journal for the Study of Greek and Latin Philosophical Traditions* 2, 2014, pp. 9–50.

F.J. Carmody, *Arabic Astronomical and Astrological Sciences in Latin Translation*, Berkeley: University of California Press, 1955.

J. Chabás and B.R. Goldstein, *The Alfonsine Tables of Toledo*, London: Springer, 2003.

Gergis, *Liber Amitegni*, MS Oxford, Bodleian Library, Laud. Misc. 594, s. XIV, ff. 142 recto–144 recto.

C.H. Haskins, *Studies in the History of Mediaeval Science*, 2nd ed., Cambridge: Harvard University Press, 1927.

Ibn al-Haytham, *Incipit liber Mamonis in astronomia, a Stephano philosopho translatus*, MS Cambrai, Médiathèque municipale de Cambrai, no. 930.

Jordanus, an International Catalogue of Mediaeval Scientific Manuscripts; online at: jordanus@ptolemaeus.badw.de

D. Juste, *Catalogus Codicum Astrologorum Latinorum*, Paris, CNRS Editions, vol. I: *Les manuscrits astrologiques latins conservés à la Bayerische Staatsbibliothek de Munich* , 2011; vol. II: *Les manuscrits astrologiques latins conservés à la Bibliothèque nationale de France à Paris*, 2015.

D. Juste, The Impact of Arabic Sources on European Astrology: Some Facts and Numbers, in *The Impact of Arabic Sciences in Europe and Asia*, Florence, 2016 (Micrologus 24), pp. 173–194.

al-Kindī and Henry Bate of Malines, *On the Revolutions of the Years of the World*, MS Vatican, *Biblioteca Apostolica Vaticana*, Pal. lat. 1407, s. XIV–XV, f. 58 recto–62 recto.

P. Kunitzsch, *Glossar der arabischen Fachausdrücke in der mittelalterlichen europäischen Astrolabliteratur*, Nachrichten der Akademie der Wissenschaften in Göttingen I, Philologisch-Historische Klasse, Jahrg. 1982, Göttingen: Vandenhoeck & Ruprecht, [1982].

Manuscripta Mediaevalia, supported by the Deutsche Forschungsgemeinschaft; online at: http://www.manuscripta-mediaevalia.de.

O. Neugebauer, *History of Ancient Mathematical Astronomy*, Berlin and New York: Birkhäuser, 1975.

O. Pedersen, The Corpus Astronomicum and the Traditions of Mediaeval Latin Astronomy, in Marian Biskup, Jerzy Bukowski, Paweł Czartoryski, et al., eds., *Colloquia Copernicana*, 4 vols., Wroclaw: Ossolineum, 1972–1975, vol. 3, pp. 57–96.

D. Pingree, review of *Hypsikles, Die Aufgangszeiten der Gestirne*, ed. and trans. Vittorio de Falco and Max Krause, *Gnomon* 40, 1968, pp. 15–16.

Pseudo-Māshā'allāh, On the Astrolabe, ed. Ron B. Thomson, Version 1.2, available online (https://shareok.org/handle/11244/14221).

al-Qabīsī, *al-Qabisi (Alcabitius): The Introduction to Astrology*, ed. and trans. C. Burnett, K. Yamamoto and M. Yano, Warburg Institute Studies and Texts 2, London: Warburg Institute, 2004.

G. Rosińska, *Scientific Writings and Astronomical Tables in Cracow: A Census of Manuscript Sources (XIVth–XVth Centuries)*, Wrocław: The Polish Academy of Sciences Press, 1984.

J. Samsó, Maslama al-Majrīṭī and the Star Table in the Treatise *De mensura astrolabii*, in M. Folkerts and R. Lorch, eds., *Sic itur ad astra: Studien zur Geschichte der Mathematik und Naturwissenschaften, Festschrift für den Arabisten Paul Kunitzsch zum 70. Geburtstag*, Wiesbaden: Otto Harrasowitz Verlag, 2000, pp. 506–22.

L. Schuba, *Die Quadriviums-Handschriften der Codices Palatini Latini in der Vatikanischen Bibliothek*, Wiesbaden: Ludwig Reichert, 1992.

F. Sezgin, *Geschichte des arabischen Schrifttums*, vol. 6: *Astronomie*, Leiden: E.J. Brill, 1978; and vol. 7: *Astrologie – Meteorologie und Verwandtes*, Leiden: E.J. Brill, 1979.

K. Takahashi, T. Mori, and Y. Kikuchihara, A Paraphrased Latin Version of Euclid's *Optica:* A Text of *De visu* in MS Add. 17368, British Library, London, *Sciamus* 3, 2002, pp. 127–92.

L. Thorndike and P. Kibre, *A Catalogue of Incipits of Mediaeval Scientific Writings in Latin*, 2nd ed., Cambridge, Mass.: Mediaeval Academy of America, 1963.

E. Zinner, *Verzeichnis der astronomischen Handschriften des deutschen Kulturgebietes*, Munich: Beck, 1925.

Warren Zev Harvey

Bernat Metge and Hasdai Crescas: A Conversation

Bernat Metge (c. 1340 – 1413) and Hasdai Crescas (c. 1340 – 1410/11) were born in Barcelona at roughly the same time. Both were bred and educated in Barcelona, and later served in the court of King Joan I and Queen Violant in Barcelona and Saragossa (1387– 1396). Metge was the foremost Catalan humanist, author of the celebrated *The Dream* (*Lo Somni*). Crescas was a distinguished rabbinical authority and author of the great anti-Aristotelian philosophic work, *The Light of the Lord* (*Or Adonai*). Metge's world was that of Cicero, Seneca, Petrarch, and Boccaccio; Crescas's was that of Rabbi Akiba, Aristotle, Averroes, and Maimonides. Metge's sources were mostly Christian and Latin; Crescas's were mostly Jewish and Hebrew.[1]

In the framework of our investigations into the agents and agency of transmission, translation, and transformation in medieval cultures, I should like to discuss the relationship between Metge and Crescas. I have subtitled my presentation "A Conversation" – not in the fashionable postmodern sense, but in the old-fashioned literal sense. Metge and Crescas in fact held face-to-face conversations. Indeed, I venture to guess that many of the transmissions, translations, and transformations in medieval cultures, as in ancient and modern cultures, took place by means of conversations, very similar to the ones that took place at the conference from which this volume is derived – in the lecture hall, in the coffee house, on the campus promenade, and elsewhere – between scholars from different countries, cultures, and disciplines.

Metge and Crescas worked together at the palace, and conversed together in Catalan. They translated and transmitted their ideas one to the other, and in so doing transformed each other's worlds. Traces of their conversations may be detected in their respective works.

1 On Metge and Crescas, see my Hasdai Crescas and Bernat Metge, pp. 141 – 154; and the Catalan version of the same article, translated and expanded (on the basis of archival material) by Jaume Riera i Sans. See also my book (in Hebrew) *Rabbi Hasdai Crescas*, pp. 82 – 87.

1 The Scribe and the Rabbi

Metge served as a royal scribe in the palace of the Crown of Aragon, and as such penned many documents concerning the kingdom's Jewish community. Crescas was recognized by the Crown as the Rabbi of all the Jews of Aragon. It is reasonable that Metge sometimes consulted with him with regard to the composition of documents concerning the Jewish community.

One Latin document, dated 25 October 1390, preserved in the Archives of the Crown of Aragon, records a meeting between the Queen, Metge, and Crescas in the royal palace in Saragossa. Crescas translated a document from Hebrew into Catalan, and, on the basis of it, the Queen issued an order to Berenguer des Cortey, her Minister of Treasury, written by her loyal scribe Metge.[2] During the anti-Jewish riots of 1391, the Queen sent an urgent letter from Saragossa, dated 18 August, to the distinguished nobleman, Pere de Queralt, in Barcelona, directing him to protect Crescas's son and companions from the mobs. The letter, preserved in the Archives of the Crown of Aragon, was written by Bernat Metge. Despite the efforts by the King and Queen, Crescas's son was murdered in the riots.[3]

2 The Soul

Metge and Crescas had similar views about the human soul. Metge discusses the question in his *Dream*, I, written in late 1398 or early 1399. After the death of King Joan in a hunting accident in 1396, Metge was sent to jail, apparently by rivals in the new regime. The book relates how the King's ghost appears to Metge in jail, and how the ghost discusses the immortality of the soul with the skeptical

2 Archives of the Crown of Aragon, Cancelleria, Reg. 2058, f. 19r–v: "mandavimus ipsum albaranum interpretari et in cathalano vulgari redigi par Atzay Cresques, judeum Barchinone expertum in talibus, quiquidem Atzay in conspecto fidelis secretarii nostri Bernardi Medici et quorundam domesticorum nostrorum albaranum predictum duxit interpretandum." Cf. the Catalan version of my Hasdai Crescas and Bernat Metge, pp. 53, 65, with the discussion of the document added by Riera i Sans.

3 Archives of the Crown of Aragon, Cancelleria, Reg. 2054, f. 102v. See Baer, *Die Juden im Christlichen Spanien*, vol. 1, pp. 676–7, no. 423: "La reyna. En P. de Queralt. Per molts e agradables serveys a nos fets per nAzday Cresques, juheu de casa nostra, nos tenim per tenguda a procurar li gracies e favors, e especial ara com li es sobiraniment necessari. ... [P]regam vox axi affectuosament, com podem, quels dits fill seu e altra companya vullats haver per recomanats e emparent los per honor nostra vullats aquells e lurs bens de tot vostre poder preservar de perills e escandols, certifficants vos, que daço farets a nos assenyalat servey, lo qual molt vos grahirem. Dada en Saragoça sots nostre segell secret... Bernardus secretariis."

scribe. Crescas's discussion of the soul is found in his *Light of the Lord*, II, 6, 1, and IIIa, 2, 1–2. The *Light of the Lord* was completed in 1410, but written over more than a decade. Both Metge and Crescas used the non-Aristotelian definition of the soul as a "spiritual substance" (*substància spiritual, eṣem ruḥani*).[4] Both rejected the Aristotelian proposition that the intellect is separate from the body, and held that it is a faculty or power of the soul (*virtut de la ànima, koaḥ ha-nefesh*).[5] Both also teach that it is the love of God – not scientific knowledge – that perfects the soul and gives it immortality.[6]

In his discussion of immortality in the Old Testament, Metge cites thirteen prooftexts: (1) Genesis 1:26 ("Let us make man in our image"), (2) Genesis 37:35 ("I will go down to *she'ol*"), (3) I Samuel 28:11–19 (the resurrection of Samuel by the Witch of Endor), (4) I Kings 17:17–23 (the revival of the son of the Shunamite by Elijah), (5) II Kings 13:20–21 (the dead man revived by Elisha's bones), (6) Psalms 16:10 ("Thou wilt not abandon me to *she'ol*"), (7) Psalms 49:16 ("God will redeem my soul from the hand of *she'ol*"), (8) Psalms 30:4 ("that I should not go down to *she'ol*"), (9) Psalms 139:2 ("Thou knowest my sitting and my resurrection"), (10) Ecclesiastes 12:7 ("and the spirit returneth unto God who gave it"), (11) Isaiah 38:10–19 ("I shall go to the gates of *she'ol*"), (12) Daniel 12:2 ("many that sleep in the dust shall awake, some to everlasting life and some to...everlasting abhorrence"), and (13) Zephaniah 3:8 ("in the day of my resurrection").[7]

Metge was not a Bible scholar. Where did he find these erudite prooftexts? Three of the above prooftexts may be traced to Christian sources: (1) Genesis 1:26, is cited similarly in Cassiodorus's *De Anima*;[8] (9) Psalms 139:2, and (13) Zephaniah 3:8, are both based on the Vulgate's homiletic translation of "*qumi*" (my rising up) as *resurrectio*. Five prooftexts may be traced to Rabbi Abraham ibn Daud's philosophic book, *The Exalted Faith* (*Ha-Emunah ha-Ramah*), written in Arabic in about 1160, and translated into Hebrew twice by two different translators connected with the circle of Crescas: (3) I Samuel 28:11–19, (7) Psalms 49:16, (10) Ecclesiastes 12:7, (11) Isaiah 38:10–19, and (12) Daniel 12:2.[9]

---

4 Metge, *Obras*, pp. 180–2; cf. *idem*, *The Dream*, p. 8. Crescas, *Or Adonai*, II, 6, 1, pp. 239, 243.

5 Metge, *Obras*, p. 190; cf. *idem*, *The Dream*, p. 11. Crescas, *Or Adonai*, II, 6, 1, p. 239.

6 Metge, *Obras*, pp. 192, 222; cf. *idem*, *The Dream*, pp. 11, 21. Crescas, *Or Adonai*, IIIa, 2, 2, p. 322.

7 Metge, *Obras*, pp. 208–10; cf. *idem*, *The Dream*, pp. 16–17.

8 Cassiodorus, *De Anima*, p. 542.

9 Ibn Daud, *Ha-Emunah ha-Ramah*, I, 7, pp. 39–41. According to Amira Eran's convincing reconstruction of events (*Me-Emunah Tamah le-Emunah Ramah*, pp. 22–25), Rabbi Solomon ibn Labi was commissioned by Crescas's close colleague Rabbi Isaac bar Sheshet (Ribash) to trans-

Four prooftexts are found in Crescas's *Light of the Lord:* (7) Psalms 49:16, (10) Ecclesiastes 12:7, (5) II Kings 13:20 – 21, and (12) Daniel 12:2.[10]

Two prooftexts are found in both the *Exalted Faith* and the *Light:* (7) Psalms 49:16, and (10) Ecclesiastes 12:7. The connection between the *Dream*, the *Exalted Faith*, and the *Light of the Lord* is seen clearly by comparing their respective uses of these two verses, as follows:

- Abraham Ibn Daud, *Exalted Faith*, I, 7 (c. 1160): "And we find that the wise Solomon, peace be upon him, says explicitly, 'And the spirit returneth unto God who gave it' [Ecclesiastes 12:7]... And his father, peace be upon him, said... He said, 'God will redeem my soul from the hand of *she'ol*, for he shall receive me' [Psalms 49:16]."[11]
- Hasdai Crescas, *Light*, IIIa, 2, 2 (1410): "[David] said, 'God will redeem my soul from the hand of *she'ol*, for he shall receive me' [Psalms 49:16]. And he said in another place... And Solomon, his son, said explicitly: 'And the spirit returneth unto God who gave it' [Ecclesiastes 12:7]."[12]
- Bernat Metge, *The Dream*, I (1398 – 9): "David...said, 'God will redeem my soul from the hand of the *infern*, for he shall receive me' [Psalms 49:16]. And [he said] in another place... Solomon, his son...said...that 'the spirit returneth unto God who gave it' [Ecclesiastes 12:7]."[13]

What are we to make of all this? How may we explain the connections between these three texts? In the first place, it must be noted that Ibn Daud's *Exalted Faith* was not translated into Latin or Catalan, and Metge did not read Arabic or Hebrew; and so he could not have read Ibn Daud's book. Similarly, he could not have read Crescas's *Light of the Lord*, even presuming an early recension of the work was available when he was writing the *Dream*.

---

late the work from Arabic into Hebrew. This translation was completed in 1392. The second translation, by Rabbi Samuel ibn Motot, a Neoplatonic mystic, was in effect a creative "emendation" of Ibn Labi's text.

10 Crescas, *Or Adonai*, pp. 319 – 20. English translation in my Ph.D. dissertation: Hasdai Crescas's Critique, pp. 479 – 83; cf. pp. 302 – 3.

11 *Ha-Emunah ha-Ramah*, 40:
ונמצא החכם שלמה ע"ה יאמר בפירוש... "והרוח תשוב אל האלוהים אשר נתנה" [קהלת יב:ז]... ואמר אביו ע"ה...
באמרו "אך אלוהים יפדה נפשי מיד שאול כי יקחני סלה" [תהלים מט:טז].

12 *Or Adonai*, 319 – 20:
ואמר [דוד] "אך אלוהים יפדה נפשי מיד שאול כי יקחני סלה" [תהלים מט:טז]. ואמר במקום אחר... ואמר שלמה בנו
בפירוש "והרוח תשוב אל האלוהים אשר נתנה" [קהלת יב:ז].

13 Metge, *Obras*, pp. 210 – 11: David...dix..."Nostre Senyor Déu reembrà la mia ànima de la mà de infern com haurà reebut mi" [Psalmi 49:16]... E en altre loch... Salamó, fill seu...dix...quel "spirit tornarà a Déu, qui ha donat aquell" [Ecclesiastes 12:7]. Cf. *idem*, *The Dream*, pp. 16 – 17.

Crescas, however, definitely did read Ibn Daud's book, and his discussion of the immortality of the soul in the Bible (*Light*, IIIa, 2, 2) is largely indebted to Ibn Daud's discussion of the subject (*Exalted Faith*, I, 7). In his discussion, he cites nine prooftexts (Genesis 5:24; II Kings 3:5; Numbers 23:10; I Samuel 25:29; Deuteronomy 30:15; Ezekiel 18:32; Psalms 115:17; Psalms 49:16; ibid. 73:17; and Ecclesiastes 12:7). With the exception of Ecclesiastes 12:7, all the verses are cited in the same order they are cited by Ibn Daud. In a parallel discussion in his polemical work, *The Refutation of the Christian Principles*, written at the same time as the *Dream*, Crescas cites ten prooftexts (Genesis 5:24; II Kings 3:5; Numbers 23:10; Psalms 142:6; 27:13; and 52:7–8; I Samuel 22:17–19; Isaiah 45:17; Exodus 19:5–6; and Deuteronomy 33:29),[14] three of which (Genesis 5:24; Numbers 23:10; II Kings 3:5) appear both in the *Exalted Faith* and the *Light of the Lord*, but none in the *Dream*.

I imagine the cited connections between the *Exalted Faith*, the *Light of the Lord*, and the *Dream* can be explained as follows. While writing the *Dream*, Metge consulted his friend Crescas about texts in the Hebrew Bible affirming the immortality of the soul. Whether or not Crescas had already written the passage in the *Light of the Lord*, IIIa, 2, 2, on the immortality of the soul, he was already impressed by Ibn Daud's discussion of the subject, and had made use of it in his *Refutation of the Christian Principles*, written at the same time as Metge's *Dream*. Crescas probably wrote down for Metge several verses, based on his memory of Ibn Daud's discussion. I am inclined to think that Crescas wrote down the verses for Metge, for otherwise it is difficult to explain the striking identity of the ways the two writers presented Psalms 49:16 and Ecclesiastes 12:7. In any case, Metge's discussion in the *Dream* of the immortality of the soul among the biblical Hebrews and Crescas's discussion in the *Light of the Lord* of the same subject testify to conversation between Metge and Crescas. In this conversation, information was transmitted (from Crescas to Metge), translated (from Hebrew to Catalan), and transformed (by Metge).

In may also be noted that in his discussion of the immortality of the soul in the Hebrew Bible, Metge makes references to Rabbinic literature which he could have heard from Crescas; e.g., the boy resurrected by Elijah is Jonah the Prophet (JT *Sukkah* 5:1; *Genesis Rabbah* 98:11, et al.), and Job is descended from Esau (BT *Baba Batra* 15a).[15]

---

14 Crescas, *Refutation*, ch. 9, pp. 76–8. The *Refutation* was written in Catalan in about 1398 but the original is lost. It survived in Rabbi Joseph ibn Shemtob's 1451 Hebrew translation: *Bittul 'Iqqare ha-Noṣrim* (see pp. 86–9 of Lasker's edition).

15 It is also possible Metge learned of Rabbinic traditions from Christian sources; e.g., the tradition identifying the resurrected boy with Jonah is found in Jerome, *Commentarium in Ionam*,

3 The Great Schism

Metge and Crescas both wrote about the Great Schism and political uncertainty. In Metge's *Dream*, II, Satan tells King Joan that he must go to Hell, for he has taken sides during the Great Schism (1378–1417), supporting the Pope in Avignon. Satan accuses all "princes of the world." They should have all remained neutral, he says, and not taken sides, since there was uncertainty as to who was the real pope. How, asks Satan, can one support an uncertain pope and thus risk denying the true pope? King Joan defends himself before Satan. To remain neutral, he argues, would necessarily mean to fail to support the true pope! The King concludes that in this situation of uncertainty the princes of the world had no choice: they had to act on the basis of their best judgment.[16]

In his *Refutation of the Principles of the Christians*, Crescas discusses the dogma of the Messiah. The Messiah, he writes, will be distinguished by wisdom and prophecy, and his coming will revive the old biblical phenomenon of prophecy. However, Jesus of Nazareth has already appeared, but prophecy has not been reinstituted. This is an argument against the messiahship of Jesus. Now, continues Crescas, if a Christian argues that prophecy is no longer necessary today, we will respond, "Yes, it is necessary today, and especially for you Christians. You need it urgently. In the days of the Bible, the prophets gave advice to kings on questions involving uncertainty. Today, the Christians are in deep uncertainty about the identity of their religious leader. Is he in Rome or in Avignon? The Christians see no way to resolve the problem. However, if there were prophecy, the prophet could resolve it for them immediately."[17]

Metge's book and Crescas's book were written at the same time (in 1398 or 1399). I fancy that the similarity of these two passages concerning the Great Schism is the result of conversations on the subject between the two, perhaps at the royal court. This was, after all, a topic hotly debated in royal courts throughout Europe. One can imagine the following conversation:

> BERNAT: The Great Schism has put the Christian kings in an impossible situation. Which pope should they support, the one in Rome or the one in Avignon? We have no way to know for certain which of them is in truth the holy man. Perhaps the kings have no choice but to act according to their conscience and best judgment, and risk error.

---

prologus, pp. 35–42: "Tradunt autem Hebraei hunc esse filium viduae Sareptanae, quem Helias propheta mortum suscitavit," etc.

16 Metge, *Obras*, pp. 238–40; cf. *idem*, *The Dream*, pp. 29–31.

17 Crescas, *Refutation*, ch. 8, pp. 65–6; Hebrew ed., pp. 75–7.

HASDAI: You are right, my friend, that the problem of the Christian kings in this matter is that they have no way to know for certain what their Christian faith demands.

BERNAT: In such a situation, could the Christian kings have any alternative but to depend on their own conscience and best judgment?

HASDAI: Yes, one can think of an alternative, and it is prophecy!

BERNAT: There is no prophecy in our days.

HASDAI: Indeed, in days of the Bible there were prophets, advisors to kings, and prophecy will be renewed in the days of the Messiah. However, it was not renewed in the days of Jesus of Nazareth, and there are no prophets today, even though the Christian kings are in need of one now urgently because of the Schism. Hey, Bernat, this is a serious argument against the Christian belief in Jesus as Messiah! I think I'll use it in the polemical tractate I'm now writing.

BERNAT: Goodness gracious, Hasdai, my friend, don't tell me now that I have helped you to think up an argument against the true Catholic faith![18]

4 Conclusion

King Joan I and Queen Violant, née Yolande de Bar, reigned from 1387 to 1396. The King, known as l'Amador de la Gentilesa, enjoyed the hunt and loved good food and fine clothes. His wife, a French duchess, was a supporter of the arts and sciences, who turned their palace in Saragossa into a cultural center for artists, writers, and Provençal troubadours, and who knew how to appreciate her two brilliant servants, Bernat Metge and Hasdai Crescas. Of his Queen, Metge wrote: "Her house was...the temple of liberality.... I do not believe that any person in the world can surpass her singular perspicacity, her intelligence, her understanding, and her boldness in undertaking great enterprises."[19]

The conversations between Metge and Crescas may be thought to have taken place primarily in the palace of the Crown of Aragon in Saragossa, perhaps sometimes in the presence of Queen Violant. In their conversations, in the Catalan language, Bernat Metge was for Hasdai Crescas a window unto classical Latin and Renaissance literature;[20] and Hasdai Crescas was for Bernat Metge a window unto Hebrew literature and medieval Aristotelian philosophy. The two

18 See my *Rabbi Hasdai Crescas*, pp. 85–7.
19 Metge, *Obras*, p. 342; cf. *idem*, *The Dream*, p. 65.
20 Since Crescas never cites Roman or Christian authors by name, it is hard to determine to what extent he exploited this window. Cf. Zonta, *Hebrew Scholasticism*, pp. 10–12.

of them, in their conversations, were agents of transmission, translation, and transformation.

Bibliography

F. Baer, *Die Juden im Christlichen Spanien*, Berlin: Akademie-Verlag, 1929.

Cassiodorus, *De Anima*, ed. James W. Halporn, *Corpus Christianorum*, Series Latina 96, Turnhout: Brepols, 1973.

Rabbi Hasdai Crescas, *Bittul 'Iqqare ha-Noṣrim*, trans. [Heb.] Rabbi Joseph Ibn Shemtob (1451), ed. Daniel J. Lasker, Ramat-Gan: Bar-Ilan University Press; Beer Sheva: Ben-Gurion University of the Negev Press, 1990.

Rabbi Hasdai Crescas, *Or Adonai*, ed. Shlomo Fisher, Jerusalem: Ramot, 1990.

Rabbi Hasdai Crescas, *The Refutation of the Christian Principles*, trans. Daniel J. Lasker, Albany: State University of New York Press, 1992.

Amira Eran, *Me-Emunah Tamah le-Emunah Ramah* [Heb.], Tel-Aviv: Ha-Kibbutz ha-Me'uḥad, 1998.

W.Z. Harvey, Hasdai Crescas and Bernat Metge on the Soul [Heb.], *Jerusalem Studies in Jewish Thought* 5, 1986, pp. 141–154 [Catalan trans. and expansion by Jaume Riera i Sans, L'ànima: un tema comú a rabí Hasday Cresques i Bernat Metge, *Calls* 4, 1990, pp. 52–81].

W.Z. Harvey, Hasdai Crescas's Critique of the Theory of the Acquired Intellect, Ph.D. diss., Columbia University, 1973 [Xerox microfilms, no. 74–1488].

W.Z. Harvey, *Rabbi Hasdai Crescas* [Heb.], Jerusalem: Zalman Shazar Center, 2010.

Rabbi Abraham ibn Daud, *Ha-Emunah ha-Ramah*, trans. Rabbi Solomon ibn Labi, ed. S. Weil, Frankfurt am Main: Typografische Anstalt, 1852.

Jerome, *Commentarium in Ionam*, ed. M. Adriaen, Corpus Christianorum, Series Latina 76, Turnhout: Brepols, 1969.

Bernat Metge, *The Dream*, trans. Richard Vernier, Burlington: Ashgate, 2002.

Bernat Metge, *Obras*, ed. Martin de Riquer, Barcelona: Universitat de Barcelona, 1959.

Bernat Metge, Written Order of Queen Violant to Berenguer des Cortey, Saragossa, 25 October 1390, Archives of the Crown of Aragon, Cancelleria, Reg. 2058, f. 19r–v.

Bernat Metge, Written Order of Queen Violant to Pere de Queralt, Saragossa, 18 August 1391, Archives of the Crown of Aragon, Cancelleria, Reg. 2054, f. 102v.

M. Zonta, *Hebrew Scholasticism in the Fifteenth Century*, Dordrecht: Springer, 2006.

Christine Chism

Transmitting the Astrolabe: Chaucer, Islamic Astronomy, and the Astrolabic Text

> But natheles suffise to the these trewe conclusions in Englisshe, as well sufficith to these noble clerkes Grekes these same conclusions in Greke. And [to] Arabiens in Arabike. And [to] Jewes in Ebrewe and to the Latyn folk in Latyn [. . .] And God woot that in alle these langages, and in many moo, han these conclusions ben suffisantly lerned and taught. And yit by diverse reules, right as diverse pathes leden diverse folke the right way to Rome.[1]

Chaucer's *Treatise on the Astrolabe* presents a pattern of transmission for astrolabic knowledge that has no single trajectory but rather is translated simultaneously to a multitude of different cultural endpoints. At these destinations, uptakes vary and different versions of astrolabic knowledge result, and yet because of the holism Chaucer attributes to the instrument itself, the ultimate body of astrolabic knowledge, like the city of Rome, nonetheless retains a world-anchoring unity amidst its regional dispersion. The pattern of transmission Chaucer describes conforms not to teleological models of knowledge transmission such as the westward tending *translatio studii*, but rather to multicentric patterns of astronomical knowledge transmission within Dār al-Islām. Astrolabic knowledge in both Islamicate and Christian cultures is produced through an intricate exploration of the astrolabe itself, which both dictates certain kinds of explorations (as users learn and exploit its capacities), and lends itself to new regional and cultural adaptations. Thus extant astrolabes from different medieval cultures become an archive of various cultural practices and values. These cultural acclimatizations create different archives of knowledge: whether for planetary mapping and the production of ephemeris, for terrestrial measurement and navigation, for trigonometric and geometrical measurement, for astrological calculation, or for religious practice (the times of prayers, the canonical hours, or the direction of Mecca from a given location on the globe). From astrolabes developed new specialized instruments – quadrants, mariner's astrolabes, and sextants – to streamline calculations of timekeeping or navigation. Writings on the astrolabe in many different languages bespeak an instrument whose innate ca-

1 Chaucer, *A Treatise on the Astrolabe*, ll. 22–26, 29–33. All citations from Chaucer's *Treatise* are taken from the *Variorum Edition* (vol. 6, *The Prose Treatises*, Part 1), edited by Sigmund Eisner (Norman, OK: University of Oklahoma Press, 2002). Citations are by line number. All citations from the *Canterbury Tales* and other Chaucerian texts are from Chaucer, *The Riverside Chaucer*, ed. Larry D. Benson, Boston, MA: Houghton Mifflin, 1987. Citations from the *Canterbury Tales* are by Fragment number (in Roman numerals) and line number.

pabilities evolve continually along with the needs of the cultures that produce them. Because the structure of the astrolabe lends itself to such a multitude of uses, we can say that the instrument itself exerts a kind of agency – suggesting new applications, and becoming a kind of mobile laboratory for adaptation and experimentation that is transmitted from culture to culture.

This paper argues that Chaucer's *Treatise on the Astrolabe* responds to the astrolabe's complex history of reinvention through transmission when it fastens precisely on the astrolabe's lability. Chaucer magnifies a sense of the astrolabe's boundless potential both as an analogue computer producing knowledge, and as a theoretical model for experimental knowledge production itself. To these ends, Chaucer's *Treatise* enacts a strikingly unassuming and non-authoritative pedagogy that vectors astrolabic knowledge to new social agencies within England: the fascinated amateurs and private experimenters who read and augment Chaucer's treatise, disseminating it in 34 highly variant and idiosyncratic manuscript versions.

1 Chaucer's Astrolabe

Chaucer begins his essay by presenting astrolabic knowledge as inexhaustible for his immediate audience: "Truste wel that alle the conclusions that han be found or ellys possible might be founde inso noble an instrument as is an astrelabie ben unknowe parfitly to eny mortal man in this regioun, as I suppose."[2] The instrument itself exceeds any transmission of knowledge about it. Thus Chaucer's *Treatise* does not pretend to complete mastery, nor can any other. This gesture sets aside textual authority and opens the door to the individual production of knowledge.

Thirty-four whole or partial manuscripts of the *Treatise* survive apparently making it Chaucer's second most popular work, after *The Canterbury Tales.*[3] These manuscripts vary in length and structure; some critics argue that conclusions 40 – 46 were added by subsequent experimenters or editors after Chaucer's death.[4] It is directed to the narrator's ten-year-old son, who is tenderly addressed both as "Lyte Lowys" and "my litel son."[5] Undoubtedly this childlike addressee

2 Chaucer, *A Treatise on the Astrolabe*, ll. 13 – 15.
3 Eagleton and Spenser, Copying and Conflation, pp. 237 – 68.
4 Eisner, (editor's intro.: Chaucer's *Treatise*), pp. 40 – 46.
5 Chaucer, *A Treatise on the Astrolabe*, l. 1; ll, 22 – 23.

stands in for a much wider readership.[6] Nevertheless, the second-person intimacy of the *Treatise*'s opening sections renders its address so personal, its directiveness so gentle, that it has also been widely treated as an early piece of children's literature.[7]

Chaucer wrote the *Treatise on the Astrolabe* around 1391, translating and compiling Latin sources attributed to Sacrobosco (d. ca. 1256) and Messahalla (= Māshā'allāh, an eighth-century Jew who may have helped plan the city of Baghdad), and drawing heavily upon Ibn al-Ṣaffār (d. 1034), a student of the eleventh-century Hispano-Arab astrologer Maslama al-Majrīṭī (d. 1007).[8] Marijane Osborne characterizes the *Treatise* as the last of Chaucer's four great translation projects, beginning with the French *Roman de la Rose*, proceeding through the Latin *Consolation of Philosophy*, and then to the wholesale rewriting of Boccaccio's Italian *Filostrato* and *Teseide*.[9] Each of these culture-crossing projects informs Chaucer's contemporaneous writing, and together they attest to Chaucer's ingestion of four major Mediterranean literary cultures: French, Latin, Italian, and, finally, Arabic.

The introduction to the *Treatise* models a similarly ecumenical and multicultural transmission, but foregrounds Chaucer's Englishness as an important endpoint among other cultural destinations:

> But natheles suffise to the these trewe conclusions in Englisshe, as well sufficith to these noble clerkes Grekes these same conclusions in Greke. And [to] Arabiens in Arabike. And [to] Jewes in Ebrewe and to the Latyn folk in Latyn, whiche Latyn folke had hem in first oute of othere dyverse langages and wroten hem in her oune tunge, that is to seyn in Latyn. And God woot that in alle these langages, and in many moo, han these conclusions ben suffisantly lerned and taught. And yit by diverse reules, right as diverse pathes leden diverse folke the right way to Rome.[10]

6 Eisner, (editor's intro.: Chaucer's *Treatise*), pp. 12–15; Laird, Chaucer and Friends, pp. 439–44.

7 Eisner, Chaucer as Teacher; and Jambeck and Jambeck, Chaucer's *Treatise on the Astrolabe:* A Handbook for the Medieval Child; Lerer (Chaucer's Sons, 909) contends that it is also a book for fathers, that teaches responsible guidance without moral didacticism

8 For a study of the extent of influence of al-Majrīṭī's astrolabe treatise on Alfonsine translations, including the *Libro del astrolabio llano*, see J. Samsó, Maslama al-Majrīṭī and the Alphonsine Book on the Construction of the Astrolabe, in *Islamic Astronomy and Medieval Spain* (Variorum Collected Studies Series, CS428), Aldershot, UK; Brookfield, Vt.: Variorum, 1994, art. XIV, pp. 1–8.

9 Osborn, *Time and the Astrolabe*, pp. 31.

10 Chaucer, *A Treatise on the Astrolabe*, ll. 22–33

Chaucer presents the Latin folk as important compilers of astrolabic knowledge from other languages who translate alongside the Greeks, Arabs, and Jews without superseding them. The various languages of Greek, Arabic, Hebrew, Latin, and English operate comparatively as conveyers for astrolabic knowledge, and they are measured against it – and against each other.[11] The *Treatise* thus effectively equalizes them – each culture receives its benefits in a calculus of well-allocated and sufficient local knowledges. This model of transmission undercuts the supersessional telos and unilinearity of the *translatio studii* and gives us instead a network of compilations existing in different languages. Chaucer renders imaginable an archive of all possible theories, descriptions, and experiments with the astrolabe, but one that manifests pragmatically and sufficiently at a variety of cultural destinations – like local branch libraries. The capacities of English in particular are at issue, because the historical association of vernacular English with the production of authorized knowledge is so recent in comparison with the illustrious *koines* of Greek, Arabic, Hebrew, and Latin. Is English sufficient to articulate and convey the knowledge to needful audiences? Chaucer answers imperatively, "Yes!" The choice of a son as audience gives this claim a forceful if baseless authority – because Father says so – and also situates the choice of language as tactical – English is sufficient for this knowledge, this act of translation and transmission, this particular familial audience. Chaucer deliberately aims his treatise low and this has stylistic ramifications. A ten-year-old reception of a technical subject will require an expository English that surrenders to its work as mediator, forgoes the rich insinuation of Chaucer's poetic writing, deliberately reducing the act of transmission to essentials to convey the knowledge as clearly and appealingly as possible.

But to work as Chaucer wishes – as an English source of astrolabic knowledge, a point in the transcultural spread of the instrument's rich, multilingual archive – the English of the treatise has to do more. And this part of the treatise sparks with a strange defensiveness about the act of translation into English. Can English operate effectively as respectable entrepot for worthy knowledge? In particular, can English stand next to Latin, the language of ecclesiastical and epistemological authority in the West, whose writers drew these same astrolabic conclusions "first oute of othere dyverese langages and wroten hem in her owne tunge"[12] during the eleventh- and twelfth-century translation movement (and not in ancient Rome as Chaucer implies)? At the end of the prologue, Chaucer

---

11 Andrew Cole (Chaucer's English Lesson) discusses the equivalence of languages here as a launchpoint for a Chaucerian theory of translation that resonates with current controversies about Wycliffite translations into plain vernacular speech.
12 Chaucer, *A Treatise on the Astrolabe*, ll. 28–29.

defensively downplays his role as merely a "lewde compilator of the labour of olde astrologiens" who has "it translatid in myn Englisshe oonly for thy doctrine."[13] Yet Chaucer had established earlier that the "Latyn folk"[14] were themselves compilers. Thus he seems to ask: "Why should English be any less capable than Latin?" The question seems to hang over him, because he then goes further to bolster the status of English, dragooning the most powerful patrons he can think of: "the king that is lorde of this langage and alle that him feithe berith and obeieth, everiche in his degre, the more and the lasse."[15] English becomes a language with an army, protected by a king. And finally, the honour of being a compiler in English is underscored with chivalric violence in the last line of the prologue: "And with this swerde shal I sle envie."[16]

The strutting defensiveness of this last line captivates me. It is the most aggressive way imaginable of saying, "I'm beneath envy, don't bother with me." Where does all this energy come from? What is at issue in being a conveyer of strange knowledge "out of othere dyverse langages"[17] into English – that it must be hedged with disclaimers, aimed low at a ten year old, and given a kingly protector with his own loyal commons? There are three possibilities. (1) The knowledge itself is somehow dangerous, perhaps because it comes ultimately from an Arabic Islamic source. In that case, it would make sense for Chaucer to hide that origin before he puts the knowledge into his young son's hands. (2) The knowledge is perfectly safe, because it has been stripped of all direct social or religious significance, made universal, and without origin – it merely renders up the terrestrial world for the user's calculation. What actually provokes anxiety is the use of the vernacular English rather than Latin to convey it. Chaucer is worried because England (and Latin Christendom itself) seems eccentric to a world of knowledge production whose centers are elsewhere – in Asia and the Mediterranean. (3) Both the knowledge and the use of English are perfectly safe – we have a positivist fantasy of the effortless translation of knowledge from culture to culture – and Chaucer's strange defensiveness is there tactically, to draw attention to the importance of the enterprise by making it sound great and dangerous. All three of these possibilities may be at work, and I don't wish to choose between them. I am more interested in how they underscore an ongoing ambiguity that attends translation and cultural transmission – the bringing of something foreign, powerful, and useful into English. Within this ambiguity, the as-

13 Chaucer, *A Treatise on the Astrolabe*, ll. 49–51.
14 Chaucer, *A Treatise on the Astrolabe*, l. 27.
15 Chaucer, *A Treatise on the Astrolabe*, ll. 45–47.
16 Chaucer, *A Treatise on the Astrolabe*, ll. 51–52.
17 Chaucer, *A Treatise on the Astrolabe*, l. 28.

trolabe exerts its instrumentalized agency, inviting to a multiplicity of perspectives, in ways that both empowered and decentered its users.

Because knowledge of the astrolabe is powerful. In Chaucer's treatise it becomes a navigational and calculating instrument designed for mobility, adaptable to different times and regions, and thus bespeaking, in its very structure, the different horizons of larger, multiple worlds. It delivers the heavens but also implicates the terrestrial world, with its rete like a spidersweb or a lady's hair net, and its shadow square, through which one can calculate the time of day from the length of a man's shadow along the ground.[18] Two capacities in particular empower the instrument. First, it is a remarkably rich and complex self-situating device. Second, it allows the operator to situate herself using multiple visible and invisible, material and conceptual frames. Because of this multiplicity, the versatility of the instrument for Chaucer seems to exert a larger figurative and conceptual potential. Marijane Osborn proposes the usefulness of what she calls "astrolabic thinking" to Chaucer's last project of the *Canterbury Tales*. Osborn demystifies the uses of this now-unfamiliar astronomical instrument and parses Chaucer's often recondite astronomical references to argue that Chaucer uses them to create temporal frames for the *Tales* that are both secular (non-astrological in its deterministic sense) and philosophically resonant.[19]

On its face, Chaucer's astrolabe is a two dimensional projection of the heavens that can be used pragmatically for a variety of different functions. Its portability and small size may limit it as a cutting-edge, precise astronomical instrument; but they render it useable at any time by anyone, on the road or at home, from an astronomical researcher interested in correcting for the precession of the equinoxes or calculating the obliquity of the ecliptic, to a ten-year-old boy. A multipurpose calendar, it gives a visual referent for the hours of the day and night, for the changing arc of visible daylight as it lengthens and shortens from summer to winter, for the seasons of the year and their celestial referents, and for the long cycle of equinoctial precession. Latin astrolabes incorporate religious markers like the saints' days, and encompasses the seven permutations

---

18 This use of shadow as clock actually happens at a key moment in the *Canterbury Tales*, though no astrolabe is in evidence, and so the time is only estimated. It is the length of Chaucer the pilgrim's own eleven-odd-foot shadow, stretched out as if for a grave, in the lengthening sunlight of the last afternoon of the pilgrimage, as the company draws near to Canterbury, and the Parson is warming up to "To knytte up al this feeste and make an ende," Chaucer, *Canterbury Tales* X, l. 47. in *The Riverside Chaucer.*

19 Osborn, *Time and the Astrolabe in the* Canterbury Tales. See also J. D. North's thorough exposition of Chaucer's cosmological knowledge, its European intertexts, and its expression – both astronomical and astrological – across his writings, in *Chaucer's Universe.*

of the year so that one can calculate the occurrence of Easter and the other moveable feasts. On a more quotidian level, one can not only pinpoint the time of day (by two kinds of hours, equal and unequal), but also when the sky will begin lightening at daybreak, and the final fading of dusk – useful, on-the-road knowledge for pilgrims and travelers. One can even enlist shadows into one's service to geometrically determine the heights of objects or the time of day if their height is known. On certain days of the year, one can measure the amount of divergence between the fixed equinoctial first point of the zodiacal sign of Aries and the actual visible constellation of Aries, and thus contemplate the vast 28,500 year cycle of the precession of the equinoxes.

The astrolabe thus puts the viewer into permutable relationships with a variety of heavenly and earthly objects, from the unimaginably distant fixed girdle of the Ptolemaic ninth sphere, to one's own shadow. Astrolabes utilize both heavenly order and heavenly contingency. Time and space are revealed as a complex dance of different cycles and fields, some grinding so slowly as to be imperceptible, and others fleeting. The astrolabe thus problematizes the absolute binary between earthly historical time and celestial transcendence, and renders possible a more complex view of many different, simultaneous temporal cycles and locations – interlaid, but differently paced.

While the astrolabe renders up a number of visible objects for use, from suns and stars to shadows and landscape objects, it also enables the operator to go beyond the sensual immediacy of everyday experience, and extrapolate from what one can see and know to what one calculates. One can know what stars are about to heave over the eastern horizon before they ascend, and during daylight; and one can know what the sun is doing as it passes on the underside of the earth. One can instrumentalize not just the Ptolomaic nesting egg of the seven planetary spheres and the eighth sphere of the stars, but also the invisible ninth sphere that drives them and drifts backward in relation to them: the Primum Mobile, girdled round with fixed and regular zodiacal signs which are not the same as the eighth sphere's visible, moving, and irregular zodiacal constellations. As such, the astrolabe is a powerful tool not only for measurement but for imaginative projection. The astrolabe at once enacts and embodies the social utility of astronomy and astrology – the rendering of the theoretical models for the movements of the heavens into the praxis of individual observation and work.

Moreover, it is both local: calibrated to provide accurate readings of celestial phenomena for particular latitudes, and global: equipped with a series of interchangeable plates adapted to differing latitudes, so that one could take it from the Levant to England and still use it simply by changing a *climate*. Because of the need for correlations with local conditions, the astrolabe is not self-contained. It was often used in conjunction with calendars and with tables – called ephemerides in English and *zīǧ* in Arabic – which charted the positions of the sun, moon, and five visible planets every day over the course of the year for a particular latitude. Chaucer is well aware of these astrolabic adjuncts, incorporating them at one point into the narrative of the *Canterbury Tales* as betokening an expertise so abstruse it verges on magic. The Franklin's Tale features a particularly deft astronomical feat (the predicting of an unusually high tide by calculating the position of the moon) and professionally outlines the array of tables and calculations necessary for the work:

> His tables Tolletanes forth he brought,
> Ful wel corrected, ne ther lakked nought,
> Neither his collect ne his expans yeeris,
> Ne his rootes, ne his othere geeris,
> As been his centris and his argumentz,
> And his proporcioneles convenientz
> For his equacions in every thyng.
> And by his eighte speere in his wirkyng
> He knew ful wel how fer Alnath was shove
> Fro the heed of thilke fixe Aries above
> That in the ninthe speere considered is.
> Ful subtilly he kalkuled al this.
> Whan he hadde founde his firste mansioun,
> He knew the remenaunt by proporcioun,
> And knew the arisyng of his moone weel,
> And in whos face, and terme, and everydeel;
> And knew ful weel the moones mansioun
> Acordaunt to his operacioun.[20]

In this passage, Chaucer reveals his currency concerning astronomical work with the celestial houses (or mansions) – derived from Islamic astrology and further elaborated by Latin Christian writers; and also the necessary adjustments resulting from the precession of the equinoxes – the divergence of the observable star Alnath (the name in medieval times for one of the "horn" stars in the constellation of Aries),[21] from the virtual "fixed" position of Aries at the start of the zodiac in the invisible ninth-sphere. The "Tolletan" or Toledan tables were astronomical tables compiled and edited in the eleventh century by Abū Isḥāq Ibrāhīm b.

20 Chaucer, *Riverside Chaucer:* V, ll. 1273–90; references to *Canterbury Tales* are by Fragment number (in Roman numerals) and line number.

21 Modern astronomers have transferred the name Alnath to one of the horn stars in Taurus; the name in Arabic *al-naṭḥ* means "butting" or "goring" with horns. My thanks to Robert Wisnovsky for this amendment.

Yaḥyā al-Zarqālluh and translated into Latin by Gerard of Cremona in the late twelfth century; they were eventually supplanted by the late thirteenth-century Alfonsine Tables, so the clerk-astronomer in the passage is a bit behind the times in this regard.[22] The Tables are well corrected because they were composed for a southern latitude and needed correcting for use in a northerly one. Chaucer's *Treatise on the Astolabe* names two friars, John Somer and Nicholas of Lynn, who had composed calendars (in 1380 and 1386 respectively), and can therefore be used to date the *Treatise* itself. In addition to tables, calendars, and other supplements, the astrolabe was often used with other instruments such as rules, quadrants, and equatorii (the latter being machines for calculating planetary movements, and thus useful for astrology); and, both in Islam and Christendom, it helped spur their further development.

This need for local adaptability in the instrument finds analogous expression in the surviving archive of astrolabe treatises across Islamic and Christian cultures. These treatises, as much as the astrolabes themselves, index the variety of uses and adaptations to which the instrument was put. In whatever language they are found – whether Arabic, Persian, Latin, Byzantine Greek, or English – they are all compilations, varying in length and content.[23] Some include instructions on how to build astrolabes; some go so far as to convey the mathematical proofs that underlie stereographic projection; and some, like Chaucer's, are pragmatic instructions for the various calculations that astrolabes of various designs can perform. Chaucer himself chooses not to convey from his source "conclusions that wol not in all thinges parformen hir bihestes"[24] for an astrolabe as small and workaday as the one he models for his son's use.

Chaucer marks the Arabic origins of his ultimate sources while situating his treatise within the compilatory culture-crossing trajectories of all astrolabic writing. Two authors have been erroneously credited with the original Arabic composition: the semi-legendary figure of the eighth-century Basran Jew, Māshā᾿allāh (the 'Messahalla' whose astrological predications determined the foundations of Baghdad), and the better documented figure of the eleventh-century Andalusian Arab, Maslama al-Majrīṭī. But it was probably al-Majrīṭī's student, Ibn al-Ṣaffār, whose treatise contributed most substantially to Chaucer's work. This attribution

22 Price, *The Equatorie of the Planetis*, p. 121; see also Chabás and Goldstein, *The Alfonsine Tables of Toledo.*

23 Edgar Laird (Geoffrey Chaucer and Other Contributors, p. 145) discusses the treatise as a *compilatio* drawing from other *compilatii:* "an assemblage, put together out of materials that were themselves assemblages, and it was, in the fifteenth and sixteenth centuries, subject to some further assembling and disassembling as it was reproduced in manuscript."

24 Chaucer, *A Treatise on the Astrolabe*, ll. 18–19.

of an ultimate (even if erroneous) Arabic source is itself revealing of the power of the linear transmission model for Western scholars.[25] One of Chaucer's immediate sources was the Latin compilation *De Compositione et Operatione Astrolabii*, which circulated from al-Andalus to England by the mid-thirteenth century, where it would have been easy for Chaucer to run across it (five or six late-medieval variant versions are still extant in Oxford and Cambridge libraries).[26] Chaucer's recognition of the compilative nature of knowledge production across transcultural networks found in astrolabic treatises is a useful corrective to more linear genealogies. Like his predecessors, Chaucer was not a passive translator, and his compilation process freely and flexibly edits his Latin source to suit his needs. Sigmund Eisner argues that Chaucer carefully adapts, and augments his treatise from many sources of astronomical knowledge – some publicly circulating, and some from his own observation – according to the perceived needs of a general audience.[27]

The astrolabe thus travels with its own technical archive, which invites further experimentation, which in turn refines the use and shape of the instrument and its adjuncts, in a kind of perpetual feedback loop. Chaucer's literary reputation comes to inform this technical generation in England. Catherine Eagleton argues that Chaucer's *Treatise* and the diagrams provided in various manuscripts influenced the design details of several extant fifteenth- and sixteenth-century astrolabes – a Y-shaped rete, letters rather than numbers around the edge, a dog's head, and a dragon's head and tail on the rete – amounting to a specifically Chaucerian astrolabe, with courtly and literate associations.[28] Similar processes of compilation and completism can be traced across the different manuscripts and recensions of Chaucer's *Treatise on the Astrolabe*.

This conception of a body of research whose parts are in continual experimental development, exchange, and competition with each other is a brilliant spur to further research and transmission of knowledge. It endows participants with a sense both of the collective force of their knowledge production, and their

25 Jenna Mead (Reading by Said's Lantern, pp. 77 – 81) cogently makes the point that a genealogy going back to Messahalla "marks" the text with the colors of Arabic and Jewish origins in order to resist its naturalization as an "unmarked white" text, to argue for a reading that resists implicit forms of orientalism. My reading diverges from hers only in that I don't think a unidirectional literary genealogy is necessary to perform such resistance.

26 Robert T. Gunther has edited and translated it, along with Chaucer's own treatise, in *Chaucer and Messahalla on the Astrolabe*. Chaucer's other main source, *De Sphaera* or *The Sphere of Johannes de Sacrobosco*, was a popular University text.

27 Eisner, (editor's intro.: Chaucer's *Treatise*), pp. 12 – 40.

28 Eagleton, "Chaucer's Own Astrolabe," pp. 303 – 25.

particular agency within it. Chaucer's own treatise brings England into this community of knowledge production and fills his readers with enthusiasm for the power and versatility of the astrolabe. Thus I am less preoccupied with the question of why Chaucer does not reveal his Arabic sources than with the shape of the model of cultural transmission he provides, which is informed by Islamicate epistemologies. As the next section discusses, the fluidity and multidirectionality of Chaucer's model of transmission actually fits the informal mechanics of knowledge production and circulation in Islamic cultures much better than the formal, teleological European genealogies of transmission, such as the *translatio studii*. And to these patterns of circulation and development, the agencies of the astrolabe and its archives are key.

2 Pragmatic Astronomy as a Multicentric Activity in Eastern and Western Islamic Cultures

Although Chaucer obscures the Arabic Islamic sources of his *Treatise*, his model of cross-cultural and multicentric knowledge production is actually a good description of the production of Islamic astronomical science as a mode of observation, pragmatic acculturation, and theoretical development across medieval Islamic societies. In Islamic regions around the Mediterranean and in the Fertile Crescent, pragmatic astronomy was arguably the most respected of all the sciences derived from non-Muslim origins, the nonreligious, *awā'il* sciences.[29] While Islamic astronomers were often obliged to justify their studies as consonant with the imperatives of Islamic culture, in relationship both to governing authorities (who could function as sporadic sources of patronage) and to the public, they also engendered widespread respect and fascination for their work. F. Jamil Ragep usefully questions the "warfare" model of the relations between sciences and religion, proposing instead historically nuanced examination of the

---

29 David Pingree (Greek Influence, pp. 32–43) discusses the culturally polygenetic sources of Islamic astronomy, which fused Greek, Indian, and Sasanian concepts. For a seminal study on translation of Greek, Sasanian, and Indian texts into Arabic, see Gutas, *Greek Thought, Arabic Culture.* See also Ahmad Dallal's useful general contextualization of Islamic scientific culture in: Science, Medicine, and Technology. Goldhizer (The Attitude of Orthodox Islam Toward the "Ancient Sciences") discusses the recurrence of anti-scientific sentiments from Islamic *faqīhs* throughout the period, while Sabra (Appropriation, pp. 230–1) points out that *faqīhs* themselves did not necessarily speak for widespread cultural assumptions, but rather exerted an influence that depended fundamentally on their situation – the influence they might or might not be able to exert on a ruler, or the popular sentiment they might or might not be able to sway.

many different ways that Islam and Hellenistic sciences such as astronomy became mutually naturalized and related through practices of "constructive engagement."[30] This section outlines some of these areas of engagement to describe an Islamic mode of scientific production that is institutionally inventive, developmentally persistent, and adaptive across regions and regimes.

David A. King differentiates the two traditions along which Islamic astronomical research and institution-building occurred, as mathematical or theoretical astronomy (*ʿilm al-hayʾa*), and folk or popular astronomy:

> To understand Muslim activity in this domain we must realize there were two main traditions of astronomy in the Islamic Near East, *folk astronomy* and *mathematical astronomy.* Folk astronomy, based on naked-eye observation of celestial phenomena and devoid of theory or computation, has generally been overlooked by historians of science with their predilection for hard core scientific achievements. Yet as we shall see, it was far more influential in Islamic society than mathematical astronomy, which as the name indicates was based on systematic observation, theory and mathematical procedures. [31]

Islamic astronomers from both camps succeeded in demonstrating the widespread utility of astronomical measurements. On the one hand, popular astronomy found some of its most intimate Islamic acculturations where royal patronage was not possible: in work associated with mosques. This led, after the eleventh century, to the development of the popular science of timekeeping or *ʿilm al-mīqāt,* which resulted in the invention of a new office associated with mosques. In addition to an imam and a muezzin, many mosques acquired a *muwaqqit,* invested with duties such as timekeeping, calculating the proper prayer times for different seasons of the year, working out the equations for crescent moon visibility (and the beginnings of important lunar months, such as Ramadan), calendar correction, and, sometimes, *qibla* direction.[32] Astrolabes were often among the instruments utilized by the *muwaqqit.* On the other hand, mathematical astronomy also succeeded in demonstrating its utility as a funding investment, as is apparent from the impressive array of observational instruments developed and invented between the ninth and fourteenth centuries: from sundials and armillaries, to mural quadrants, parallactic rulers, enormous gnomons and azimuth rings.

Early on, Islamic astronomers developed a wide array of specialized subfields in which to nurture expertise. This emphasis on specialization promoted

---

30 Ragep, Freeing Astronomy from Philosophy, at pp. 49 and 62.
31 King, Science in the service of religion, Art. I, p. 246 in *Astronomy in Service of Islam.*
32 King, On the Role of the Muezzin and the *Muwaqqit*, pp. 285–346.

collaboration and the continual rechecking of each other's observations and calculations. The observatory was an institutional bastion of this professionalization; it brought together dynamic communities of intellectuals among the *ʿulamāʾ* who continually adapted to changing funding conditions or travelled to find others in new locations. Aydın Sayılı emphasizes that astronomical observatories were generally short-lived in Islamic culture, unlike all the other Islamic signatory institutions: the madrasa/university, the hospital, and the public library.[33] As institutions, observatories could not rely on long-term patronage; and, because of their uncertain consonance with Islamic religious practice, observatories were not normalized as a recipient of *waqf* revenues – except in extraordinary cases such as the Maragha observatory, which lasted an unprecedented 40 years.

At other times, however, well supported, and well equipped observatories – such as those at Maragha (late thirteenth century), Samarkand, and Istanbul – could amass libraries and teaching support, and, essentially, become research institutions for promulgating astronomical knowledge. A. I. Sabra draws attention to the important problem presented to cultural historians by the ongoing active support of Islamic cultures for the rational sciences, which made their way into *madrasas* (where they were not officially on the curriculum); for it is clear that – across many regions and periods – interest was strong and ongoing enough to ensure a continuity of research and dialogue between scientific, mathematical, and astronomical scholars.[34]

Astronomical treatises circulated from established observatories and their libraries, such as the early ninth-century Shammasiya observatory in Baghdad or the thirteenth-century Maragha observatory in Azerbaijan, which boasted a library of 40,000 books.[35] Collaborative research practices were nurtured by the

33 Sayılı, *The Observatory in Islam*, pp. 4–6.

34 Sabra, Appropriation, pp. 223–43. Sabra outlines a three-stage process – acquisition; scientific, Hellenistic development; and Islamic enculturation – by which the Greek sciences were actively seized and naturalized to become an intrinsic part of Islamic cultures. His article usefully questions the dubious assumptions underlying the idea of Islamic "mediation" of science between the Greeks and the Latins. Edward S. Kennedy (Al-Bīrūnī's Treatise on Astrological Lots, pp. 9–54) translates and comments on an astrological treatise of al-Bīrūnī's that argues for the validity of an Indian source describing eclipses, to convince a patron that the Greeks are not the only astronomers whose models have validity.

35 Kennedy (Late Medieval Planetary Theory, pp. 365–78) discusses the development of a school of planetary theories at Maragha where Naṣīr al-Dīn al-Ṭūsī, Quṭb al-Dīn al-Shīrāzī, and contemporary astronomers from regions as far as China and the Maghrib worked in tandem, producing work that eventually found its way into Copernicus's writings on the orbits of the moon and Mercury.

culture of field specialization, leading to research staffs working together on observation projects in order to improve planetary charts and gain better measurements for ecliptic inclination, the shift of the solar apogee, and the rate of equinoctial precession. The need for patronage begat fierce competition between thinkers and scientists, and fostered critical thinking and methodological stringency.[36] For instance, a fascinating letter from Jamshīd al-Kāshī gives a lively picture of the scientific community of fifteenth-century Samarkand, nurtured and participated in by its Mongol patron: Ulugh Beg.[37] Observatories could thus be public or private, state-sanctioned or independent, huge or tiny; but they supplied the means by which communities of scholars could come together collaboratively (in person, or in manuscript form) to innovate.

Thus, uncertain funding did not doom the institution of the observatory and the practice of observational astronomy in Islam, but rather honed requisite acculturation skills. Islamic observatories are characterized by fluid response to conditions of intermittent patronage support, resident expertise, and popular enthusiasm. Mathematical astronomy and judicial astrology, together with the related popular arts of timekeeping and calendar correction, made astronomical practices vital for a number of cultural endeavors. Even in periods without sustained aristocratic or royal support, there was considerable room for individual displays of initiative – such as Sayyid Rukn al-Dīn's astonishing Institute of the Time and the Hours (Mu'assasa Waqt wa Sa'āt completed in 1325 in Yazd) and the possible private observatories of Ibn Yūnus (d. 1009) in Cairo, and Ibn Bājja of al-Andalus (d. 1139).[38]

Pragmatic astronomy, with its many subfields and specialists, developed recognizable "laboratory practices" and methodological imperatives, including an ongoing investment in empirical measurement, and the continual invention and refinement of observational instruments. Theoretical astronomy also persisted, but often independently from the institutional support of pragmatic initiatives.[39] Some of the most provocative theoretical astronomers, such as the

---

36 For a good case study of opportunistic adaption, in the quest for knowledge, to historical circumstances (such as the Mongol invasions), see *Naṣīr al-Dīn al-Ṭūsī's Memoir on Astronomy*, vol. 1, pp. 3–20.

37 Mohammad Bagheri translates it from the Persian in A Newly Found Letter of Al-Kāshī on Scientific Life in Samarkand, pp. 241–56. See also Kennedy, Ulugh Beg as Scientist; and *idem*, The Heritage of Ulugh Beg.

38 For a vivid description of the Institute of the Time and the Hour, with its heliotropic copper bird, enormous water clock, wheel shaped mechanical timer, and a discussion of the indeterminacies of the institute's uses as a curiosity, *muwaqqit* or observatory, see Sayili, *The Observatory in Islam*, 236–44.

39 Sayılı, *The Observatory in Islam*, p. 250.

Damascene Ibn al-Shāṭir and most of the Western astronomers of al-Andalus, were not associated with institutionalized observatories. Yet, as Sabra has noted, Andalusian scientific culture emphasizes unusual independence of thought and intellectual ambition, as well as a sense of Andalusian exceptionalism. Andalusian astronomers such as Ibn Rushd and al-Biṭrūjī freely questioned Ptolemaic astronomical models or synthesized them with Aristotelian modes of thought. By contrast, eastern and Asian astronomers, such as Ibn al-Haytham and the Maragha astronomers, tended to work within Ptolemaic models, refining them with inventions such as the Ṭūsī couple, in order to bring even the problematic Ptolemaic equant hypothesis into better accord with observed phenomena.[40]

This practice of multiplex institutional self-maintenance among both mathematical and popular astronomers proved very resilient because of its sheer outreach and deep cultural penetration. Islamic astronomy not only provided a form of cultural capital to the rulers who intermittently patronized it, but also provoked a widespread interest among the *ʿulamāʾ* and the astrologically enthusiastic populace. Widespread support for astronomy persisted despite the fact that astronomical observation for scientific research and more accurate modeling of the cosmos was not always separable from the Islamically dubious practice of judicial astrology.[41] Because astrology, like philosophy, was decried by influential Islamic theologians such as al-Ghazālī, astronomers could suffer the vagaries of a patron like the eleventh-century ruler of Cairo, al-Ḥākim. al-Ḥākim sporadically patronized astronomical research, frantically sought astrological consultation throughout his rule, and half-built an observatory (only to lose interest) – yet in 404 AH went so far as to banish astronomers and forbid celestial observation. However, on a systemic level, astronomical practice often prospered from its association with astrology; rulers patronized astronomers, funded their observations and their larger scale instruments, and constructed observatories for them, while successfully demanding astrological measurements and treatises even from those who disapproved of judicial astrology. And a sideline in astrology could incite local support from the populace at large.[42]

---

40 See Sabra, The Andalusian Revolt, pp. 133 – 153; *idem*, Configuring the Universe, pp. 288 – 330; and Kennedy, Late Medieval Planetary Theory.

41 However, methodological arguments for their separation were made by figures as influential as Ibn Sīnā, whose work was invoked repeatedly until it became a standard view; see: Ragep and Ragep, The Astronomical and Cosmological Works of Ibn Sīnā, pp. 3 – 15.

42 Two articles that illuminate both the social aversion to, and social utility of, Islamic judicial astrology are: Saliba, The Role of the Astrologer, pp. 45 – 67; and Burnett, Al-Kindī on Judicial Astrology, pp. 77 – 117.

Given this history of the acculturation of Islamic astronomy, George Saliba takes issue with historiographies that trace a post-eleventh-century decline of Islamic science; he points out that (1) the golden age might have been the age of royal patronage, but royal patronage was unreliable at the best of times; and (2) Islamic astronomy achieved its most notable theoretical and practical breakthroughs after the eleventh century.[43] In the East, astronomical advances are evidenced in the work of Ibn al-Haytham, Mu'ayyad al-Dīn al-ʿUrḍī (d. 1266), Naṣīr al-Dīn al-Ṭūsī (d. 1274),[44] Quṭb al-Dīn al-Shīrāzī (d. 1311), Ibn al-Shāṭir al-Dimashqī (d. 1375), and Ṣadr al-Sharīʿa al-Thānī (d. 1347).[45] In the West, problems of Ptolemaic astronomy were tested and reformulated by Ibn Bājja of Saragossa (d. 1139), Ibn Ṭufayl of Granada (d. 1185/6), Ibn Rushd of Cordova (1126–1198), al-Biṭrūjī, probably of Seville (1200) and a student of Ibn Ṭufayl, and Jābir b. Aflaḥ of Seville (1200).

Saliba posits that Islamic astronomy actually prospered more after its liberation from palace patronage systems, because it had successfully infiltrated the everyday life of the mosque and its communities, become accessible to laymen, and won acceptance even from religious imams.[46] Because Islamic astronomers continually had to prove their social utility, they pioneered a scientific practice that was socially responsive, discovering more ways of proving useful as the ongoing support of rulers remained uncertain. The loosening of patronage systems and noble courts necessitated more individual and collaborative research. This need for collaboration is responsible for a generalization of knowledge production into a shifting but persistent network, rather than the genealogical line of a central foundation, its illustrious patrons, and its ongoing institutional self-maintenance.

---

43 Saliba, *A History of Arabic Astronomy*, p. 65; see also King (The Astronomy of the Mamluks, pp. 531–55) on the achievements of Mamluk astronomers.

44 F. Jamil Ragep (Tusi and Copernicus, pp. 145–63) argues that Copernicus's heliocentrism is preconditioned by models for the earth's rotation found in al-Ṭūsī's discussion of comets – one prominent attestation of a discussion of the earth's rotation that extends for 600 years and surfaces in Shīrāzī, Jurjānī, Qūshjī and Bīrjandī. In addition to Copernicus's adaptation of equant-solving innovations such as the Ṭūsī couple, these borrowings suggest that Copernicus is better read "within the tradition of late medieval Islamic astronomy, more so than that of medieval Latin scholasticism" (ibid., 145). See also Ragep, ʿAlī Qushjī and Regiomantanus, pp. 359–71; and *idem*, Copernicus and his Islamic Predecessors, pp. 65–81.

45 For Ṣadr al-Sharīʿa's persistent refinements to the solutions for Ptolemaic problems devised by Ṭūsī and Shīrāzī, see *An Islamic Response to Greek Astronomy*, Ahmad S. Dallal's edition of the *Kitāb Taʿdīl Hay'at al-Aflāk*.

46 Saliba, *A History of Arabic Astronomy*, p. 65.

Islamic astronomy also profited from a cultural respect for the empirical and the observable that was widespread throughout Islamic cultures. Unlike medieval Christianity, Islam rarely tried to establish itself as contrafactual and miraculous. In Islamic cultures, eyewitness observation has its own authority and was incorporated into the testimonial aspects of religious practice; the reliability of the *isnāds* were tested – and that focus on observational testing and research ended up bolstering the authority of the *ḥadīth.* George Saliba highlights "the Islamic philosophical tradition – in which truth was supposed to be within a system that is consistent, harmonious, and well articulated, with religion having an essential position in that system."[47] Because of this cultural respect for consistency within and between disciplines, there was widespread receptivity to the idea that astronomy was consonant with Islamic religious knowledge – astronomy coexisted with, and supplemented theological truths without impinging on them. Even al-Ghazālī exempted astronomy from his attack on the foreign and *awā'il* sciences, because of the obvious observational truth of astronomical phenomena such as eclipses.[48]

The sheer number of astronomical and astrological manuscripts attested from a multitude of sites proves the force and persistence of both mathematical and folk astronomy as modes of knowledge within different Islamic societies. These manuscripts range from technical manuals such as astrolabe treatises,[49] to mathematical proofs of the principles underlying new developments in astronomical instrumentation,[50] catalogues of the stars and their major constellation groupings,[51] ephemerides and tables,[52] theoretical treatises correcting and developing new theorems for the movements of planets and satellites within Ptolemaic astronomy,[53] explications of judicial astrology, [54] and new equations calculating the conditions under which the crescent moon can be visible.

---

47 Ibid., 62.

48 al-Ghazālī, *al-Munqidh min al-ḍalāl*, pp. 68 – 69.

49 See King, *Al-Khwarizmi*; Berggren, Abu Sahl al-Kuhi's Treatise, pp. 141 – 252; al-Zarqālluh, *al-Shakkāziyya*; and Ibn Bāṣo, *Risālat al-ṣafīḥa*.

50 al-Farghānī, *On the Astrolabe.*

51 Al-Ṣūfī's catalogue of stars, the *Kitāb ṣuwar al-kawākib*, informed the Alfonsine *IIII Libros de la Ochaua Espera* (Four Books on the eighth sphere) and subsequent compilations; see Samsó and Comes, Al-Sufi and Alfonso X, pp. 67 – 76.

52 See Kunitzsch, *The Arabs and the Stars.*

53 See Kennedy, Late Medieval Planetary Theory; Ragep, ʿAlī Qushjī and Regiomantanus; and *idem*, Copernicus and his Islamic Predecessors.

54 Edward S. Kennedy's collection, *Astronomy and Astrology in the Medieval Islamic World*, shows how methodologically interrelated were problems in astrological and astronomical calculations – the same mathematical and trigonometric skills were needed for both.

In sum, we can gather an image of Islamic astronomy as a loose and continually reforming multicentric network, where no resource was stable but where the continual dissemination of texts and local traditions of observational practice created a productive cultural medium. New research centers, observational habituses, and scholarly genealogies were generated and retransmitted among the peripatetic scholars of the Muslim world. Ultimately, the writers of the astronomical and astrological treatises from which Chaucer eventually drew accord to his description of multicentric transmission: the gathering of knowledge from many languages and its redirection into many cultural destinations. Chaucer's emphasis on pragmatic experimentation – his editing away of what is not useful in his source – also echoes back to the Islamic astronomers' careful redaction, rethinking, and critical departures from their primary Greek astronomical sources: Aristotle and Ptolemy.

3 The Astrolabe – Transmission and Acculturation

The multicentricity of Islamic astronomical science finds both an index point and an engine in the astrolabe itself. We can use astrolabes and astrolabe treatises to mark particular developments in, and widespread popularization of, astronomical knowledge as it acculturated to the everyday lives of their users. In Islamic cultures, astrolabes were employed primarily for what we could call astrology and folk astronomy: the practical "applications of astronomy to daily life in the Muslim community."[55] As an instrument adapted to the everyday utilities that knit celestial observation to Islamic religious practice, the astrolabe was, in its Muslim popular origins, both an acculturated instrument and a multifunctional one. It functioned not as the experimental cutting edge of Islamic mathematical astronomy and knowledge production, but rather as a hinge that rendered science socially relevant, translating astronomy into society. Along with various other instruments and astronomical tables, it was linked to the pragmatic problems of astronomical calculation that preoccupied the muezzins, and, after the seventh/thirteenth century, the *muwaqqit*s of the religious institution.[56] Islamic astrolabes varied greatly according to region, but often included quadrants on the back of the mater for calculating the *qibla*, and for determining as-

55 King, preface to *Astronomy in the Service of Islam*, p. xiv.
56 See King, On the Role of the Muezzin and the *Muwaqqit*. For a more recent series of studies on timekeeping, see King, *In Synchrony with the Heavens*.

tronomically defined times for prayer throughout the day on the shadow square, as well as for trigonometric calculations of size, direction, and distance.[57]

While the principles of stereographic projection and early devices based on them were developed in ancient Greece and the antique Mediterranean by the fourth century, Morrison argues that the development of the astrolabe as an instrument took place in the Near East by the year 900. John Philoponus of Alexandria describes a workable astrolabe by the sixth century, and pre-Islamic Persian and Levantine astrolabes existed by the seventh century. The first Muslim credited with making an astrolabe (ca. 750) is Muḥammad ibn Ibrahīm al-Fazārī (d. 796 or 806), an astronomer associated with the ʿAbbāsid court of al-Manṣūr, and a translator of Sanskrit astronomical works into Arabic.[58] Islamic astrolabes generally comprised a "mother" plate (*umm*) with a marked raised rim, or "limb;" several climate plates, each containing a stereographic projection of the heavens at a particular latitude; a rete (a netlike rotating plate containing a projection of certain key fixed stars and a circle of the ecliptic); a suspension from which to hold the astrolabe vertically; and an alidade and pointer to align with heavenly objects.

Over the next four centuries the astrolabe was refined and adapted as it was disseminated throughout Islamicate cultures from al-Andalus to India – and into Europe in the twelfth and thirteenth centuries.[59] The back of the astrolabe in particular was a space for refinements: azimuth curves, twilight arcs, shadow scales, sine quadrants, prayer-time scales, azimuths of the *qibla*, canonical hours scales, calendars for calculating the date of Easter in a given year, and longitude scales.[60] However, there was also considerable variation in which pointer stars are chosen for the rete.[61] David A. King suggests that extant astrolabes rep-

---

57 Morrison, *The Astrolabe*, p. 14. For a survey of the sheer variety of Islamic astrolabes see Pingree, *Eastern Astrolabes*.

58 Morrison, *The Astrolabe*, p. 31–49. Early Islamic culture was profoundly influenced by Sasanian writings, and later by Persian writings, which themselves were informed by Indian traditions; so Indian influences are also traceable in Islamic astronomical texts. See Pingree, Indian Influence, pp. 118–26.

59 David A. King (The Neglected Astrolabe, pp. 45–55) discusses Islamic refinements and developments to the astrolabe, with reference to contemporary textual sources about them.

60 Morrison, *The Astrolabe*, p. 37. The 1388 astrolabe of Jaʿfar b. ʿUmar b. Dawlatshāh exemplifies the way in which astrolabes themselves became intercultural archives. It includes eight semicircles in the lower half of the back including Dorotheus's names of zodiacal signs and the lords of the decans, both Greek (indicated by last letters of their name) and Indian (indicated by symbols); see Pingree, *Eastern Astrolabes*, pp. 42–3.

61 Paul Kunitzsch (Al-Ṣūfī and the Astrolabe Stars, pp. 151–166) discusses the variation between al-Ṣūfī's treatises on which stars (from a total of fifty-five) might be chosen for the rete.

resent goldmines of historical and archival information: Islamic astrolabes were often signed by their makers and precisely situated through particular climate latitudes – thus counterposing geographical, biographical, linguistic, and artistic/cultural data in their archives of inscription.[62]

Astrolabic treatises accompanied and drove the dissemination of the instrument. As early as the ninth century, Abū al-ʿAbbās Aḥmad b. Muḥammad b. Kathīr al-Farghānī (fl. ca. 820-after 861) wrote a theoretical and pragmatic *Book on the Construction of the Astrolabe*, outlining the fundamental mathematical principles behind astrolabic construction (such as the preservation of circles), in order to rescue the instrument from merely traditionalized regional modes of construction, and to provide the irrefutable "proof of its correctness" and "evidence of its theoretical basis" that had been lacking.[63] Astrolabe treatises were written in Persia by al-Bīrūnī (973–ca. 1048), Naṣīr al-Dīn al-Ṭūsī (1201–1274), and ʿAbd al-Raḥmān b. ʿUmar al-Ṣūfī (987–7).[64]

Astrolabic knowledge entered Europe via southern France, and possibly Mediterranean trade, though that trajectory is harder to trace.[65] Arianna Borrelli's recent definitive study illuminates the plurality of meanings and uses ascribed to the astrolabe by its earliest tenth- and eleventh-century Latin adopters; the astrolabe is refracted into a "loose collection of co-existing 'aspects'" through many divergent sources: extant astrolabe artifacts, texts, geometrical diagrams, and illustrations.[66] Borrelli argues persuasively that the astrolabe was so provocative an instrument for European scholars because it could function simultaneously as a conceptual model, a sublimation of complex mathematical concepts into more easily imagined geometrical figures, and a practical instrument.[67]

Chaucer's ultimate sources came later, and included John of Seville's mid-twelfth-century translation into Latin of Ibn al-Ṣaffār's treatise on the astrolabe, falsely attributed to Māshāʾallāh: *De compositione et utilitate astrolabii.*[68] Gerbert of Aurillac (ca. 945–1003), who became Pope Sylvester II (999–1003), recommended the astrolabe's utility for monastic canonical timekeeping, but did not

62 King, Bringing Astronomical Instruments Back to Earth, pp. 3–53.

63 al-Farghānī, *On the astrolabe*, pp. 24–5.

64 David A. King surveys the Islamic manuscripts that mythologize or investigate the origins of astrolabes, both etymologically and etiologically: King, The Origin of the Astrolabe According to Medieval Islamic Sources.

65 Morrison, *The Astrolabe*, 37.

66 Borrelli, *Aspects of the Astrolabe*, p. 18.

67 Ibid., p. 33.

68 Morrison, *The Astrolabe*, p. 37.

recommend it for laymen.[69] Adelard of Bath (ca. 1080-ca. 1160) – who learned Arabic and Islamic science on his travels throughout the Mediterranean – wrote an astrolabe treatise: *De opere astrolapsus*, dedicated to the nephew (*nepos*) of Henry I of England (most probably Henry Plantagenet – the future Henry II – but possibly Henry of Blois).[70] More so than Chaucer's treatise, Adelard's tone is pedagogically condescending, directed to a child in need of basic explanations of the Ptolemaic cosmos: its sphericity, the obliquity of the ecliptic, and the climates. The treatise describes the instrument's parts before proceeding to the use of the instrument, beginning with its utility as a surveying instrument using the shadow square, continuing to its astronomical and navigational functions (which reveal some large mistakes about the latitudes of particular cities), and then finally detailing its time-telling and astrological capacities.[71] Adelard also translated the *Zīj* tables of al-Khwārizmī, and transmitted principles of geometry and trigonometry by translating, from Arabic, Euclid's *Elements* and information on sines. His own work, *Quaestiones naturales*, provided explanations for a wide variety of natural phenomena.[72] Hermann Contractus (1012–1054) compiled a treatise on the astrolabe in Germany, based on manuscripts translated from Arabic by the assiduous translation workshop at the monastery of Santa Maria de Ripoll.[73] By the fourteenth century in England, Chaucer's two main sources, both technical manuals: the *De Sphaera* of Sacrobosco and *De compositione et utilitate astrolabii*, had circulated widely and were readily available.

By the end of the thirteenth century, the astrolabe was accepted throughout Europe as a tool for basic astronomy, which itself changed radically from devotional exercise to science as the wealth of Islamic astronomical and astrological treatises entered Europe through translation movements in al-Andalus and the south of France. Morrison underscores the astrolabe's galvanizing influence on the whole conception of astronomical science in Europe:

> Introduction of Islamic astronomy into Europe was revolutionary [...] there was virtually no European appreciation of geometric astronomical constructions or analytic procedures for predicting future astronomical events. Observations of celestial phenomena were crude and derivative. European astronomical knowledge in this era was more literary and philosoph-

---

69 Ibid. He also devised a new kind of abacus as a teaching tool for the quadrivium; see Burnett, *The Introduction of Arabic Learning into England*, pp. 10–12.

70 Burnett, The Education of Henry II, pp. 31–60.

71 Cochrane, *Adelard of Bath*, pp. 97–106. Adelard also translated one of the most influential Arabic astrological treatises; see Abū Maʿshar, *Abbreviation of the Introduction to Astrology, together with the Medieval Latin Translation of Adelard of Bath.*

72 Cochrane, *Adelard of Bath*, p. 1.

73 Morrison, *The Astrolabe*, pp. 37–8.

> ical than scientific and astronomical discussion primarily celebrated the glory of God's creation. Islamic astronomy was built on geometrical models justified by observation. The astrolabe introduced both of these scientific approaches to Europe. The astrolabe was not the only element that began the process of establishing astronomy as a science in the Latin West, but it was arguably the most significant.[74]

Between the tenth and seventeenth centuries in both Europe and the Near East, astrolabes were refined and adapted; surviving instruments attest to enormous regional and occupational variability throughout Europe, and an even greater variability throughout Eastern and Western Islamic regions. There was a significant amount of parallel evolution – and independent reinvention – with regard to refinements, and with regard to subsidiary instruments such as quadrants. A little-known universal time keeping device, the *navicula de Tenetiis*, underwent parallel development in ninth-century Baghdad and fourteenth-century England.[75] A universal astrolabe was developed independently several times, most notably around 1048 in Toledo by ʿAlī ibn Khalaf and Arzachel (al-Zarqālluh) (1028–87), who each describe a different mechanism. Al-Zarqālluh's latitude-independent astrolabe became known in Europe as the Saphea Arzachelis. In Louvain, Belgium, Gemma Frisius (1508–1555) drew upon the Saphea and possibly other texts for his influential *de Astrolabo Catholicum* (1550). Ibn Khalaf devised what became known as the *lamina universalis*, which was reinvented by John Blasgrave of England, in the "Mathematical Jewell" (1585).[76]

European astronomers and instrument producers did not, however, introduce any astronomical or navigational uses that had not already been developed in the Near East and Mediterranean three to five centuries earlier. The history of the astrolabe is thus a history of reinvention and continual refinement with only discontinuous dissemination between regions and cultures. It parallels the development of Islamic astronomy itself: dynamic, situational, opportunistic, and multifocal. This history suggests that the astrolabe seems never to have presented itself as a closed book to its users and developers – there was always room for new discovery. The dissemination of these discoveries (or lack thereof) bespeaks the complexity of regional and occupational technological networks. Aiding the astrolabe's spread was its double cachet as an instrument that was at once esteemed and popularizable. The unusual numbers of astrolabes still extant from the ʿAbbāsid period attest to the prestige accorded a well-constructed instrument, but it was also obtainable in cheaper models for the use of ama-

74 Ibid.
75 King, 14th-Century England or 9th-Century Baghdad?, pp. 204–26.
76 Morrison, *The Astrolabe*, 169–90.

teurs, enthusiasts, and functionaries. Some astrolabe workshops sold paper printouts of the plates and projections which could be mounted on wood and assembled into a working instrument by the user.

The professionalization of the *muwaqqit* as a functionary in mosques after the thirteenth century,[77] also impelled technological refinements. These refinements would eventually lead to the astrolabe's replacement, at least for purposes of timekeeping. After about 1300, astrolabes began to be supplanted in Islamic cultures by less expensive and less elaborate quadrants, which, though less multifunctional, were easier to use for telling time. However, the astrolabe in Islamic cultures held on for much longer than it did in Europe; in particular regions, astrolabes continued to be made and used up through the nineteenth century and the onset of colonialism.[78]

By comparison, the redevelopment of the astrolabe in Europe is belated, but shows an almost equally extensive range of adaptation and variation across regions.[79] After astrolabes were transmitted to medieval Latin and Byzantine Christian regions, the back of the astrolabe in particular was adapted – the shadow square was used for canonical hours, not Muslim prayer times, and new calendars were devised for calculating the date of Easter in a given year. The astrolabe's utility as a timekeeping instrument made it particularly attractive in monasteries; this accelerated its reacculturation, although it did not entirely supplant traditional methods for figuring out the proper time to rise and sing the canonical hour of Matins, such as calibrated candles, or timed psalms.[80] European astrolabes were refined and adapted by individual users and workshops up through the seventeenth century. After that time, the astrolabe was supplanted by more accurate astronomical, navigational, and timekeeping devices, such as telescopes, sextants, and mechanical clocks.[81]

In sum, the astrolabe, wherever we find it, was tied into the everyday life and cultural practices within the regions and communities in which it was used. And, as a layman's instrument – separate from, or only supplemental to, the learned traditions of mathematical astronomy – it operated in Muslim and Christian cultures with the everyday cultural utility that we might associate with a cell

---

77 King, On the Role of the Muezzin and the *Muwaqqit*.

78 Morrison, *The Astrolabe*, p. 32.

79 Roderick and Marjorie Webster (*Western Astrolabes*) survey a wide variety of Western instruments, revealing the large range of individual refinements and choices.

80 Morrison, *The Astrolabe*, p. 40.

81 Francis M. Rogers (*Precision Astrolabe*) traces the long afterlife of the artificial-horizon mariner's astrolabe, which was refined into the artificial-horizon sextant and used by Portuguese aviators in the "second age of adventure," in the early twentieth century.

phone – personally modified apps and all: providing instant solutions to astronomical, timekeeping, and navigational problems. I would argue that the astrolabe both metaphorizes and impels Islamicate cultures of pragmatic applied astronomy. Thinking about Chaucer's astrolabe in the context of Islamic applied astronomy reminds us that the mechanical is never just the mechanical, and a tool is never just a tool. In this, the astrolabe's history of instrumental variance, acculturation, and redevelopment accords broadly with Hans Rheinberger's analysis of the productive dynamics of techno-experimental systems:

> Scientists are, first and foremost, *bricoleurs* (tinkerers), not engineers [...] established tools can acquire new functions in the process of their reproduction. Their insertion into a productive or consumptive process beyond their intended use may reveal characters other than the original functions they were designed to perform.[82]

Drawing upon Gaston Bachelard's notion of "phenomenotechnique" – the idea that "technology is not just an ancillary result of scientific investigation but also the very modus operandi of science"[83] – Rheinberger explores how observational instruments exert agency in the production of knowledge. He argues that scientific concepts are inextricable from scientific process – the interactions of instruments, laboratory procedure, and observed phenomena. Technological instrumentation shapes the data it produces, and thus becomes a conceptual agent in a complex and mutually informing process: "Phenomenon and instrument, object and experience, concept and method are all engaged in a running process of mutual instruction."[84] With its history of constant, pragmatic adaptation, the astrolabe's open-ended utility was the crux of its allure because its operators could determine for their own purposes what forms of knowledge it would yield.

In sum, the astrolabe's Islamic history of individual experimentation and cultural adaptation foregrounds Chaucer's unusual treatment of it as at once: (1) a powerful scientific instrument, (2) a cognitive mode, and (3) a model for a different kind of pedagogy that sidelines the replication of authority to foster instead hands-on experience, observation, and experiment. I would argue that in Chaucer's *Treatise on the Astrolabe* we have a rare opportunity to examine this instrumental agency in the medieval world, and further to see how it transgresses our own post-medieval divisions between science, literature, and pedagogy.

---

82 Rheinberger, *Towards a History of Epistemic Things*, p. 32.

83 See Lenoir's foreword (Epistemology Historicized, p. xii) to Rheinberger's *An Epistemology of the Concrete*.

84 Ibid., p. xiii.

As Chaucer encapsulates the properties and experimental potentials of the astrolabe with deliberate lack of exhaustiveness, his treatise encourages further research, and coaxes a fascination with the multiformity into which the instrument can shape observable phenomena. When the narrator of the *Treatise* points out that the capacities of the astrolabe overreach local sufficiencies of knowledge, he allures his addressee with the open-ended mystery of ongoing exploration. Chaucer's future reader, invoked in the immature body of his son, is thus invited to situate himself – in multiple different exercises – within a defamiliarized, fathomless, but nontheless marvelously calculable universe. Although Chaucer's treatise subjects the future stargazer to a "certain nombre of conclusions,"[85] they cannot exhaust the capacities of the instrument and the inventiveness of its individual user and compiler.

This sense of inexhaustible capacity makes Chaucer's *Treatise* inspiring in three significant ways. It transmits not only interest in, and fascination with, the instrument, but also a more fundamental interest in the processes of observation and experimentation themselves. Finally, in its positioning as an instructional treatise for a little boy, it functions within a pedagogy of empowerment and liberation. It is with this pedagogy, and its role in inciting a culture of amateur exploration extending throughout the fifteenth century in England, that I will now conclude.

4 The Powers of Pedagogy in the Fifteenth Century

As many critics have noted, *A Treatise on the Astrolabe* is a model of carefully humanized technical explanation. In Part 1, which describes the astrolabe's parts and uses, Chaucer employs vivid metaphorical language, comparing the radiating azimuths to a spider's claws, and the rete, a filigreed star map, to a lady's hair net. In Part 2, in the midst of more abstract and challenging calculations, he will suddenly invoke the wonder of a stargazer's experiential encounter with a night sky, "in som wynters night whan the firmament is cler and thikke sterred."[86]

More provocatively, *The Treatise on the Astrolabe* brings strange, difficult, and prodigious learning close to home: its title is editorial but four of its early manuscripts call it "*bred and mylk for children.*" In describing its parts, Chaucer

85 Chaucer, *A Treatise on the Astrolabe*, l. 10.
86 Chaucer, *A Treatise on the Astrolabe*, ll. 833–34.

accentuates the essential maternity of the instrument, translating Latin *mater* (Arabic *umm*), the rimmed base of the instrument as "moder" (mother) and giving it a womb: "[T]he moder of thin astrelabye is thikkest plate perced with a large hool that resceiveth in hir wombe the thynne plates compowned for diverse clymates and thy reet shapen in manere of a nett or of a webbe of a loppe."[87] This maternal imagery, I would argue, reminds the reader of how productive and fecund the instrument has been set up to be. In addition to signaling the astrolabe's fecundity, the maternal and feminine language familiarizes the alien instrument: It becomes part of a family-narrative between father, instrumentalized mother, and son that renders even the *Treatise*'s most abstract experiments homely. [88]This provides a starting place through which Chaucer can lead from the familiar to the strange, and back again. For at the same time, the familiar is invested with a striking newness. Use of the astrolabe amounts to a wholesale reconceptualization of everyday phenomena. A shadow can become a clock; the night sky can be a calculator. Other geographies truly have other horizons, and one can calculate what they will look like, pinpoint the unique specificities of one's own, and come to a new view of the geography shared with other Englishmen as one geography of many – not center, not only home, but one of many possible homes from which one can see the same stars slightly differently.

By interlarding the everyday, limited, and local with the wonderful, excessive, and unknown, the narrator invigorates the everyday with the wonders of new frames and perceptions. *The Treatise of the Astrolabe* thus reverses what John Fyler has called the domestication of the exotic, in his reading of the Squire's Tale, a domestication so complete that it sucks its own narrative dry of energy, turning it into a hung-over and torpid version of the Same.[89] By contrast, *The Treatise on the Astrolabe* first lures the child into its intricacies by making then homely, but then goes on to defamiliarize the entire world.[90]

Even more interestingly, in *The Treatise* the pleasure the narrator takes in this pedagogical process begins to unmoor the structural hierarchies of the

87 Chaucer, *A Treatise on the Astrolabe*, ll. 96–99. Jenna Mead (Reading by Said's Lantern, p. 81) draws attention to this maternal imagery.
88 Seth Lerer (Chaucer's Sons) proposes a more paternalistic model for this pedagogy, placing it in the Macrobean tradition.
89 Fyler, Domesticating the Exotic, pp. 1–26.
90 Patricia Claire Ingham (Little Nothings: The Squire's Tale and the Ambition of Gadgets) discusses how technical gadgets such as astrolabes instill a pleasure and wonder in the new that spurs human invention and helps reenchant the material world in Chaucer's Squire's Tale and medieval romance more generally.

teacher-student relationship. It does so, by invoking discourses of friendship.[91] In the opening to *The Treatise*, the narrator transforms the training process into a network of pleasurable exchange between friends: [92]

> Lyte Lowys my sone I aperceyve wel by certeyne evidences thyn abilite to lerne sciences touching nombres and proporciouns. As wel considre I thy besy praier in special to lerne the tretys of the astrolabie. Than for as moch as a philosofre saiht: he wrappith him in his frende that condescendith to the rightfull praiers of his [frende], therefore have I yoven the a suffisant astrolabie as for oure orizonte compowned after the latitude of Oxenforde, upon which by mediacioun of this litel tretys, I purpose to teche the a certein nombre of conclusions aperteyning to the same instrument.[93]

This opening is pedagogically predictable when it constructs the student as eagerly longing to learn and the instructor as "condescending" to teach, while tracking the development of a student's innate capacities. However, the proverb from the "philosofre,"[94] which has not yet been traced, oversets this hierarchy. Since when, in medieval pedagogical relationships, are teacher and student – let alone parent and child – *friends?* George Lyman Kittridge finds "friend" an inappropriate mode of address when directed to one's own son, though one might so address the son of a friend and colleague.[95] Larry D. Benson finds this word inappropriate as a mode of address from a mature man to a ten-year old child. His textual commentary concludes that Chaucer is merely translating his source, whatever it was.[96] The treatise itself changes tone from condescension to empowerment as it shifts from the preceptorial first part to the business-like second part. However, as the proverb surprisingly indicates, the teacher does not simply endow instruction; rather he is swaddled, "wrapping himself" in his student/kinsman/friend, in a relationship which connotes both equality and mutual – but not quite commensurable – pleasure. The child gets the pleasure of the learning he apparently desires, the parent gets the pleasure of "wrapping" – surrounding himself by, being enfolded within – the child. The *Treatise on the As-*

---

91 Edgar Laird usefully discusses and presents analogues to the theme of friendship in astrolabe treatises and other technical writings (Laird, Chaucer and his Friends).

92 In the notes to his edition, Skeat suggests some loose rendering from Cicero's *De amicitia* about the need to accede only to the honorable requests of ones friends. But what comes across in Chaucer's phrase is not wariness but warmth.

93 Chaucer, *A Treatise on the Astrolabe*, ll. 1–11.

94 Chaucer, *A Treatise on the Astrolabe*, l. 5.

95 Kittredge, Lewis Clifford or Lewis Chaucer?, speculates that the Lewis to whom Chaucer directs the treatise is not his own son but actually a son of Lewis Clifford, chamber knight to Richard II and friend of Chaucer; he finds "friend" a more natural epithet in that context.

96 *The Riverside Chaucer*, p. 1095, n. 6.

trolabe thus acknowledges the hierarchies implicit in pedagogical transmission of knowledge but seeks to use them for mutual pleasure – a companionship of equals that produces not only knowledge but friendship.

The shape of Chaucer's treatise as it works out (as opposed to its original stated plan) further accentuates the child's agency in this learning process. Chaucer's plan (like many of his unrealized literary plans) was huge and ambitious. Part 1 was to describe the components of the astrolabe and give the Arabic- and Latin-derived terminology of each part. Part 2 was to give exercises appropriate for a small portable astrolabe. Part 3 was to contain many tables; Edgar Laird characterizes Chaucer's list as "breathless," giving "the impression that Chaucer had not completely thought out just what tables he would include but intended to have lots of them."[97] The tables were to include a catalogue of fixed stars along with their latitudes and longitudes, a table of the declensions of the suns, tables for the longitudes of cities and towns, tables for the "governance of a clokke"[98] to better find the altitude meridian, and more exercises that could be done using the calendars of Friar John Somer and Friar Nicholas of Lynn. Part 4 was to include a practical account of the movement of celestial bodies, such as the sun, moon, and planets, along with a table to show, hour by hour, the movements of the moon every day, and the methods to know in which degree of the zodiac the moon will rise in any latitude, and the arising of any planet through its latitude from the ecliptic line. All this is knowledge that would be helpful not only for astrological calculation but also for travel – why else include calculations for the moon's rise in any latitude? Finally Part 5 would be a general introduction to wider astrological theory that had been transmitted to Europe "after the statutes of oure doctors," [99] along with more tables for the equations of the astrological houses (for the latitude of Oxford), tables of the dignities of planets, and various other details.

This massive plan would have been an enormous work of calculation, collation, and translation – equipping the child at once with practical astronomy, navigation, timekeeping, and astrology. It would have been pedagogically overwhelming, though an amazing resource for a burgeoning culture of amateur astronomers and astrologers – many of whom in the fifteenth century calculated, compiled, and trafficked in these materials. Chaucer's *Treatise* is frequently com-

97 Laird, Geoffrey Chaucer and Other Contributors, p. 150; he also mentions (pp. 150–1) that some manuscripts containing Chaucer's *Treatise* themselves include other astronomical works and tables perhaps in response to Chaucer's uncompleted plans – I would argue they seem to respond even more to the expansiveness of his enthusiasm.

98 Chaucer, *A Treatise on the Astrolabe*, ll. 65–66.

99 Chaucer, *A Treatise on the Astrolabe*, ll. 79–80.

piled with other related astronomical and astrological works, as though to provide fuller coverage of its plan, and Catherine Eagleton demonstrates through stemmatic analysis that *Treatise* scribes went out of their way to consult other manuscripts of the *Treatise* to render it more complete.[100] However, most manuscripts end the *Treatise* with Part 2: the immediate list of astrolabic exercises that anyone could perform with a small astrolabe and a small number of already circulating tables. This ending transforms the *Treatise* from a planned field-defining compendium into a workable practicum that doesn't lead to any particular pedagogic end; the variant number of conclusions across the diverse manuscripts attest to its modularity. Each exercise is headed by a statement of which conclusion can be reached; the user of the astrolabe can choose between them by scanning across titles, and new exercises can always be added.

The treatise as a whole introduces the user to a large amount of technical vocabulary, much of it wearing its Arabic origins on its face, from the almucanters and azimuth lines on the plates, to the names of the stars to be used as orientation points. In addition, I would argue, Arabic underlies the treatise in unstated ways, emerging in one of Chaucer's most homely metaphors: the web of a lop (spider) for the rete, whose name in Arabic was *ʿankabūt* (spider). Chaucer knew this because one of his Latin sources carefully gives the Arabic word and translates it. On the one hand, the rete does look like a spider's web or a multilegged spider; but on the other hand, it is interesting to think of how what is linguistically alien underlies what positions itself as most familiar and even playful – calculated to appeal to a child's need for visually imaginative analogues.

Chaucer's energetic redirection of his sources to the purposes of imaginative pedagogy become even clearer if we compare his treatise with an almost contemporaneous astrolabe treatise: the earliest astrolabe treatise written in French, by Pèlerin de Prusse, produced in 1362 within the court of Charles V, for an audience that was not as mathematically equipped as Chaucer's but very interested in astrology. Pèlerin's mode of address is businesslike rather than figural, and he establishes a stronger pedagogical telos for it. While covering many of the same exercises as Chaucer and drawing on the same sources, Pèlerin offers his treatise first and foremost as an astrological manual for learning to cast horoscopes. When he gets to a discussion of the astrological twelve houses, he directs the

100 Eagleton, Copying and Conflation, p. 259.

reader to pay unusual attention: "Ce chapitre es le plus hault et plus commun et souvent neccessaire sur toutes autres."[101]

By contrast, the *Treatise* refuses to situate the reader on any single path. Instead it gives the child (and the novice adult readers for which he stands in) an instrument through which they can situate themselves within a calculable universe. Although it subjects the child to an unspecified number of conclusions, they are open-ended ones, expanded from Chaucer's sources, and augmented by *The Treatise*'s many scribes. The manuscript history of the *Treatise* comes to dramatize Chaucer's reluctance ever to fasten upon just one authoritative end. Even the overdetermined ending of *Troilus and Criseyde* doesn't have *forty-six* conclusions.

While clear causalities are problematic – and I am not proposing that Chaucer's treatise triggered a revolution in Western observational astronomy – the treatise's widespread dissemination as the second most popular of Chaucer's works suggests that Chaucer's pedagogical choices to accentuate the *Treatise*'s utility and easy uptake widened his treatise's appeal. It fed a growing interest in observational science; and one that did, in fact, gather steam in the fifteenth century, which also saw – not coincidentally – the beginning of a spread of European astrolabe production apexing in the late fifteenth and sixteenth centuries, and even a specifically Chaucerian line of astrolabes.[102]

It is virtually impossible to find a general history of Western science that does not dismiss the medieval period or condescend to Islamic scientific discoveries – so powerful is the telos for modern scientific discovery, with its high points of observation, experimentation, replicability of results, and institutional dissemination and reward. Sabra critiques the deformations enjoined by both "precursorism" (to Western science) and preserverism (of Greek science) on discussions of Islamic science that ought to be situated and historicized as productive practices within Islamic cultures.[103] Happily, recent work on the material processes by which science is actually produced to serve particular historical

101 "This chapter is the most important and most often necessary of all the chapters" (*Pèlerin de Prusse on the Astrolabe*, pp. 52–3).

102 See Park, Observation in the Margins. Roderick and Marjorie Webster (*Western Astrolabes*, p. 8) list notable fifteenth- and sixteenth-century astrolabe makers ranging throughout Europe: in Paris, Jean Fusoris (1365–1436); in Louvain, Gualterus Arsenius (1554–1579); and in Nuremberg, Georg Hartmann (1489–1564). For Chaucerian astrolabes see Eagleton, "Chaucer's Own Astrolabe."

103 Sabra, Appropriation, pp. 223–43; see also King, Some Reflections, pp. 1–5.

and cultural situations can perhaps open a different, less teleologically skewed picture.[104]

The pragmatics of art and science between many Islamic cultures and many Latin Christian ones illuminate centuries of volatile cultural production – from the eighth to the eighteenth century – that defy traditional epochal periodizations. Perhaps that resistance to periodization is the very thing that makes so elusive a quantifiable Islamic "contribution" to the "progress" of astronomy as world science. We look in vain for evidence of the massive paradigm shifts that are the foundation gestures of such narratives, for a Galileo or a Newton. What we get instead is a culture of ongoing, persistent, and widespread scientific and empirical pragmatism, with many theoretical advances, many conclusions, no single telos. Yet that culture in itself, I would argue, was productive enough to provoke admiration, emulation, and replication in the European medieval and early modern periods. The great twelfth-century translation moment in Spain and south France responded to the riches discovered in the large Andalusian archives and libraries captured during the conquest of Iberia, while cultures of Islamic scientific patronage was eagerly emulated by Spanish and later European rulers.[105] George Saliba and David Lindberg describe how the gradual European "discovery" of Arabic literate culture helped impel scholarly investigation from the medieval through the early modern periods.[106] And the allure of Islamic science as cultural production was not channeled only toward Europe, a prelude to narratives of the rise of European hegemony, but rather provoked responses in many of the cultures touched by Islam – it is as multifocal as Chaucer claimed.[107]

Chaucer's *Treatise on the Astrolabe* with its complex back-histories of multicultural compilation and multifocal transmission can offer a useful corrective to the precursorism and preserverism of that traditional narrative. In addition, I would argue, the treatise not only transmits a sense of other histories and other practicums in its content, but also encourages ongoing experiment and amateur exploration in its pedagogical stance: of friendship and pleasure in

104 For a recent revisionist foray that makes a beginning, see Kwa, *Styles of Knowing.*

105 Borrelli (*Aspects of the Astrolabe*, pp. 20–72) theorizes oral as well as textual and instrumental modes of transmission: the teaching, discourse, and pedagogical exchange characterizing many cultural practices in Andalusian and south Italian contact zones.

106 Saliba, *A History of Arabic Astronomy*; Lindberg, *The Beginnings of Western Science*, pp. 215–44. See also chap. 2 of Edward Grant's *The Foundations of Science in the Middle Ages* (pp. 18–32), entitled: "The New Beginning: The Age of Translation in the Twelfth and Thirteenth Centuries."

107 For one example of Indian uptake – which is as culturally filtered and kinematic as that in Europe, see Pingree, Indian Reception, pp. 471–85.

learning and teaching (bearing in mind that "amateur" etymologically derives from "love"). Ultimately, Chaucer's treatise fosters an experimental culture imagined as multiple exchanges of knowledge, observational power, and pleasure, and one that is abetted and metaphorized by the multifunctionality and cultural adaptability of the astrolabe itself. And that may have been the most transformative transmission of all.

Bibliography

Abū Maʿshar, *Abbreviation of the Introduction to Astrology, together with the Medieval Latin Translation of Adelard of Bath*, ed. and transl. Charles Burnett, Keiji Yamamoto, and Michio Yano, Leiden: E. J. Brill, 1994.

M. Bagheri, A Newly Found Letter of Al-Kāshī on Scientific Life in Samarkand, *Historia Mathematica* 24, 1997, pp. 241–56.

J.L. Berggren, Abu Sahl al-Kuhi's Treatise on the Construction of the Astrolabe with Proof: Text, Translation and Commentary, *Physis* 31, 1994, pp. 141–252.

A. Borrelli, *Aspects of the Astrolabe: "Architectonica ratio" in Tenth- and Eleventh-Century Europe*, Stuttgart: Franz Steiner Verlag, 2008.

C. Burnett, Al-Kindī on Judicial Astrology: "The Forty Chapters," *Arabic Sciences and Philosophy* 3, 1993, pp. 77–117.

C. Burnett, The Education of Henry II, in *The Introduction of Arabic Learning into England*, London: The British Library, 1997, pp. 31–60.

C. Burnett, *The Introduction of Arabic Learning into England*, London: The British Library, 1997.

J. Chabás and B.R. Goldstein, *The Alfonsine Tables of Toledo*, Dordrecht, Boston, London: Kluwer Academic Publishers, 2003.

Geoffrey Chaucer, *A Treatise on the Astrolabe*, ed. Sigmund Eisner (*A Variorum Edition of the Works of Geoffrey Chaucer*, vol. 6, *The Prose Treatises*, Part 1), Norman: University of Oklahoma Press, 2002.

Geoffrey Chaucer, *Canterbury Tales*, in *The Riverside Chaucer*, ed. Larry D. Benson. Oxford: Oxford University Press, 1987.

L. Cochrane, *Adelard of Bath, The First English Scientist*, London: British Museum Press, 1994, pp. 97–106.

A. Cole, Chaucer's English Lesson. *Speculum* 77/4, 2002, pp. 1128–67.

A. Dallal, Science, Medicine, and Technology: The Making of a Scientific Culture, in John L. Esposito, ed., *The Oxford History of Islam*, Oxford: Oxford University Press, 1999.

C. Eagleton, "Chaucer's own Astrolabe:" Text, Image and Object. *Studies in History and Philosophy of Science* 38, 2007, pp. 303–25.

C. Eagleton and M. Spenser. Copying and Conflation in Geoffrey Chaucer's *Treatise on the Astrolabe:* a Stemmatic Analysis Using Phylogenetic Software. *Studies in History and Philosophy of Science* 37, 2006, pp. 237–68.

S. Eisner, Chaucer as Teacher. *Children's Literature Association Quarterly* 23.1, 1998, pp. 35–39.

al-Farghānī and R. Lorch, ed. and transl., *On the astrolabe: Arabic text edited with translation and commentary by Richard Lorch*, Stuttgart: Steiner, 2005.

J. Fyler, Domesticating the Exotic in the Squire's Tale, *English Literary History* 55, 1988, pp. 1–26.

Abū Ḥāmid al-Ghazālī, *al-Munqidh min al-ḍalāl*, Cairo: Anglo-Egyptian Press, 1964.

I. Goldhizer, The Attitude of Orthodox Islam Toward the "Ancient Sciences," in Merlin L. Swartz, transl. and ed., *Studies in Islam,* New York and Oxford: Oxford University Press, 1981, 184–213.

E. Grant, *The Foundations of Science in the Middle Ages: Their Religious, Institutional, and Intellectual Contexts*, Cambridge: Cambridge University Press, 1996.

R.T. Gunther, ed. and transl., *Chaucer and Messahalla on the Astrolabe*, Early Science at Oxford 5, Oxford: Oxford University Press, 1929.

D. Gutas, *Greek Thought, Arabic Culture: The Graeco-Arabic Translation Movement in Baghdad and Early 'Abbasid Society: 2nd–4th/8th–10th Centuries,* London: Routledge, 1998.

S. Horobin,The Scribe of Bodleian Library MS Bodley 619 and the Circulation of Chaucer's Treatise on the Astrolabe, *Studies in the Age of Chaucer* 31, 2009, pp. 109–124.

Abū ʿAlī al-Ḥusayn Ibn Bāṣo and E. Calvo Labarta, ed. and transl., *Risālat al-ṣafīḥa al ŷāmiʿa li-ŷamīʿ al-ʿurūḍ* = Tratado sobre la lámina general para todas las latitudes; edición crítica, traducción y estudio por Emilia Calvo Labarta, Madrid: Consejo Superior de Investigaciones Científicas, Instituto de Cooperación con el Mundo Árabe, 1993.

P. C. Ingham, Little Nothings: Chaucer's Squire's Tale and the Ambition of Gadgets, *Studies in the Age of Chaucer* 31, 2009, 53–80.

T. J. Jambeck and K. K. Jambeck, Chaucer's *Treatise on the Astrolabe:* a Handbook for the Medieval Child. *Children's Literature* 3, 1974, pp. 117–22.

E.S. Kennedy, Al-Bīrūnī's Treatise on Astrological Lots, in *Astronomy and Astrology in the Medieval Islamic World*, Aldershot: Ashgate, 1998, art. XV, pp. 9–54.

E.S. Kennedy, The Heritage of Ulugh Beg, in *Astronomy and Astrology in the Medieval Islamic World*, Aldershot: Ashgate, 1998, art. XI, pp. 1–12.

E.S. Kennedy, Late Medieval Planetary Theory, *Isis* 57/3, 1966, pp. 365–78.

E.S. Kennedy, Ulugh Beg as Scientist, in *Astronomy and Astrology in the Medieval Islamic World*, Aldershot: Ashgate, 1998, art. X, pp. 1–6.

D.A. King, 14th-Century England or 9th-Century Baghdad? New Insights on the Elusive Astronomical Instrument Called *Navicula de Venetiis*, *Centaurus* 45, 2003, pp. 204–26.

D.A. King, Bringing Astronomical Instruments Back to Earth: the Geographical Data on Medieval Astrolabes (to ca. 1100), in Arjo Vanderjagt & Lodi Nauta, eds., *Between Demonstration and Imagination: Essays in the History of Science and Philosophy Presented to John D. North*, Leiden and Boston: Brill, 1999, pp. 3–53.

D.A. King, *Al-Khwarizmi and New Trends in Mathematical Astronomy in the Ninth Century* (Occasional Papers on the Near East 2), New York: Hagop Kevorkian Center for Near Eastern Studies, 1983.

D.A. King, *Astronomy in the service of Islam*, Aldershot, Hampshire, Great Britain; Brookfield, Vt., USA: Variorum, 1993.

D.A. King, The Astronomy of the Mamluks, in *Islamic Mathematical Astronomy*, London: Variorum Reprints, 1986, art. III, pp. 531–55.

D.A. King, *In Synchrony with the Heavens: Studies in Astronomical Timekeeping and Instrumentation in Medieval Islamic Civilization*, vol. 1: *The Call of the Muezzin (Studies I–IX)*, Leiden: Brill, 2004.

D.A. King, The Neglected Astrolabe, in Menso Folkerts, ed., *Mathematische Probleme im Mittelalter: Der lateinische und arabische Sprachbereich,* Wiesbaden: Harrassowitz Verlag, 1996, pp. 45–55.

D.A. King, On the Role of the Muezzin and the *Muwaqqit* in Medieval Islamic Society, in F. Jamil Ragep and Sally P. Ragep, with Steven Livesey, eds., *Tradition, Transmission, and Transformation: Proceedings of Two Conferences on Pre-Modern Science Held at the University of Oklahoma*, Leiden and New York: E.J. Brill, 1996, pp. 285–346.

D.A. King, Science in the service of religion: The case of Islam, in *Astronomy in the service of Islam*, Aldershot, Hampshire, Great Britain; Brookfield, Vt., USA: Variorum, 1993, art. I, pp. 245–262.

D.A. King, Some Reflections on the History of Islamic Astronomy, in *Islamic Mathematical Astronomy,* London: Variorum Reprints, 1986, art. I, pp. 1–5.

G. Kittredge, Lewis Clifford or Lewis Chaucer. *Modern Philology* 14.9, 1917, pp. 513–518.

P. Kunitzsch, Al-Ṣūfī and the Astrolabe Stars, in Fuat Sezgin, ed., *Zeitschrift für Geschichte der Arabisch-Islamischen Wissnschaften*, Frankfurt: Institut für Geschichte der Arabisch-Islamischen Wissenschaften, 1990, pp. 151–166.

P. Kunitzsch, *The Arabs and the Stars: Texts and Traditions on the Fixed Stars and their Influence in Medieval Europe*, Northampton: Variorum Reprints, 1989.

C. Kwa, *Styles of Knowing: A New History of Science from Ancient Times to the Present*, Pittsburg: University of Pittsburg Press, 2011.

E. Laird, Chaucer and Friends: The Audience for the *Treatise on the Astrolabe. Chaucer Review* 41/4, 2007, pp. 439–44.

E. Laird, Geoffrey Chaucer and Other Contributors to the *Treatise on the Astrolabe*, in Thomas Prendergast and Barbara Kline, eds., *Rewriting Chaucer: Culture, Authority and the Idea of the Authentic Text, 1400–1602*, Columbus: Ohio State University, 1999, pp. 145–65.

T. Lenoir, Epistemology Historicized: Making Epistemic Things, foreword to Hans-Jorg Rheinberger, *An Epistemology of the Concrete: Twentieth-Century Histories of Life*, Durham: Duke University Press, 2010.

S. Lerer, Chaucer's Sons, *University of Toronto Quarterly* 73/3, 2004, pp. 906–16.

D.C. Lindberg, *The Beginnings of Western Science: The European Scientific Tradition in Philosophical, Religious and Institutional Context, 600 B.C. to A.D. 1450*, Chicago and London: University of Chicago Press, 1992, pp. 215–44.

J. Mead, Geoffrey Chaucer's *Treatise on the Astrolabe. Literature Compass* 3/5, 2006, pp. 973–91.

J. Mead, Reading by Said's Lantern: Orientalism and Chaucer's *Treatise on the Astrolabe*, *Al-Masaq* 15/1, March 2003, pp. 77–82.

J.E. Morrison, *The Astrolabe*, Rehoboth Beach: Janus, 2007.

J.D. North, *Chaucer's Universe*, Oxford: Clarendon Press, 1988.

M. Osborn, *Time and the Astrolabe in the* Canterbury Tales, Norman: University of Oklahoma Press, 2002.

K. Park, Observation in the Margins, 500–1500, in Lorraine Daston and Elizabeth Lunbeck, eds., *Histories of Scientific Observation*, Chicago and London: University of Chicago Press, 2011.

Pèlerin de Prusse, *Pèlerin de Prusse on the Astrolabe: Text and Translation of his* Practique de Astrolabe, ed. and transl. Edgar Laird and Robert Fischer, Binghampton: Medieval and Renaissance Texts and Studies, 1995.
D. Pingree, *Eastern Astrolabes, Historic Scientific Instruments of the Adler Planetarium and Astronomy Museum*, vol. 2, Chicago: Adler Planetarium and Astronomy Museum, 2009.
D. Pingree, The Greek Influence on Early Islamic Mathematical Astronomy, *Journal of the American Oriental Society* 93/1, 1973, pp. 32–43.
D. Pingree, Indian Influence on Sasanian and Early Islamic Astronomy and Astrology, *The Journal of Oriental Research: Special Volume Dedicated to H. H. Sri Chandrasekharendra Sarasvati*, ed. V. Raghavan, Madras: Kuppuswami Sastri Research Institute, 1973, pp. 118–26.
D. Pingree, Indian Reception of Muslim Versions of Ptolemaic Astronomy, in F. Jamil Ragep and Sally P. Ragep, with Steven Livesey, eds., *Tradition, Transmission, and Transformation: Proceedings of Two Conferences on Pre-Modern Science Held at the University of Oklahoma*, Leiden and New York: E.J. Brill, 1996, pp. 471–85.
D.J. Price, *The Equatorie of the Planetis, edited from Peterhouse Ms. 75.1*, Cambridge: Cambridge University Press, 1955.
F.J. Ragep, ʿAlī Qushjī and Regiomantanus: Eccentric Transformations and Copernican Revolutions, *JHA* 34, 2005, pp. 359–71.
F.J. Ragep, Copernicus and his Islamic Predecessors: Some Historical Remarks, *History of Science* 45, 2007, pp. 65–81.
F.J. Ragep, Freeing Astronomy from Philosophy: An Aspect of Islamic Influence on Science, *Osiris*, 2nd series 16, Science in Theistic Contexts: Cognitive Dimension, 2001, pp. 49–64, 66–71.
F.J. Ragep, Tusi and Copernicus: The Earth's Motion in Context, *Science in Context* 14/1–2, 2001, pp. 145–63.
F.J. Ragep and S.P. Ragep, The Astronomical and Cosmological Works of Ibn Sīnā: Some Preliminary Remarks, in Nasrollah Pourjavady and Živa Vesel, eds., *Sciences, Techniques et Instruments dans le Monde Iranien (Xe – XIXe Siècle)*, Tehran: Institute Français de Recherce en Iran, Presses Universitaire d'Iran, 2004, pp. 3–15.
H.-J. Rheinberger, *Towards a History of Epistemic Things: Synthesizing Protein in the Test Tube*, Stanford: Stanford University Press, 1997.
F.M. Rogers, *Precision Astrolabe: Portuguese Navigators and Transoceanic Aviation*, Lisbon: Dilsar; Taunton, MA: Wm. S. Sullworld, 1971.
A.I. Sabra, The Andalusian Revolt against Ptolemaic astronomy, in Everett Mendelsohn, ed., *Transformation and Tradition in the Sciences: Essays in Honor of I. Bernard Cohen*, Cambridge: Cambridge University Press, 1984, pp. 133–153.
A.I. Sabra, The Appropriation and Subsequent Naturalization of Greek Science in Medieval Islam: A Preliminary Statement, *History of Science* 25, 1987, pp. 223–43.
A.I. Sabra, Configuring the Universe: Aporetic, Problem Solving, and Kinematic Modeling as Themes of Arabic Astronomy, *Perspectives on Science* 6/3, 1998, pp. 288–330.
Ṣadr al-Sharīʿa al-Thānī, *An Islamic Response to Greek Astronomy: Kitāb Taʿdīl Hayʾat al-Aflāk of Ṣadr al-Sharīʿa*, ed. Ahmad S. Dallal, Leiden and New York: E.J. Brill, 1995.
G. Saliba, *A History of Arabic Astronomy*, New York: NYU Press, 1994.
G. Saliba, The Role of the Astrologer in Medieval Islamic Society, *Bulletin D'Études Orientales: Sciences Occultes et Islam* 44, 1992, pp. 45–67.

J. Samsó, Maslama al-Majrīṭī and the Alphonsine Book on the Construction of the Astrolabe, in *Islamic Astronomy and Medieval Spain* (Variorum Collected Studies Series, CS428), Aldershot, UK; Brookfield, Vt.: Variorum, 1994, art. XIV, pp. 1–8.
J. Samsó and M. Comes, Al-Sufi and Alfonso X, in *Islamic Astronomy and Medieval Spain* (Variorum Collected Studies Series, CS428), Aldershot, UK; Brookfield, Vt.: Variorum, 1994, art. XVII, pp. 67–76.
A. Sayılı, *The Observatory in Islam and its Place in the General History of the Observatory*, Ankara: Türk Tarih Kurumu, 1960.
Naṣīr al-Dīn al-Ṭūsī, *Naṣīr al-Dīn al-Ṭūsī's Memoir on Astronomy (= al-Tadhkira fī 'ilm al-hay'a)*, ed. and comment. F.J. Ragep, 2 vols., New York: Springer-Verlag, 1993.
R. Webster and M. Webster, *Western Astrolabes*, Historic Scientific Instruments of the Adler Planetari, Chicago: Adler Planetarium & Astronomy Museum, 1998.
Ibn al-Naqqāsh al-Zarqālluh (al-Zarqālī) and Roser Puig, ed. and transl., *Al-Shakkāziyya: Ibn al-Naqqāsh al-Zarqālluh, Edición, traducción y estudio por R. Puig,* Barcelona: Instituo "Millas Vallicrosa" de Historia de la Ciencia Árabe, 1988.

Frank T. Coulson

Literary criticism in the Vulgate Commentary on Ovid's *Metamorphoses*

Of the numerous Latin commentaries written on the *Metamorphoses* in the High Middle Ages, perhaps none was so influential as that commentary generally referred to as the "Vulgate."[1] First, the circulation of the text itself was wide-ranging, both in its geographical dissemination, as well as the number of manuscript copies produced.[2] Secondly, although the Vulgate Commentary was not printed, its approach to the explication of the poem has been shown to have influenced the later Renaissance commentary of Raphael Regius, who is considered to be the most important humanist commentator on the *Metamorphoses.*[3] And thirdly, the Vulgate Commentary may be considered a seminal work in the evolution of the Latin commentary tradition on Ovid because of its highly sophisticated approach to the explication of the text. Unlike the purely allegorical tradition represented by such texts as Arnulf of Orléans's *Allegoriae*[4] and Pierre Bersuire's *Ovidius moralizatus*,[5] the Vulgate Commentary adopts a multifaceted approach, and demonstrates a highly developed interest not only in questions of grammar and syntax, but also in what may be termed "modern" interpretive questions, such as structure, characterization, style, and Ovidian influence on subsequent writers. In the present article, I consider a more discrete aspect of the approach

1 I have written on the Vulgate Commentary elsewhere. See my Ovid's transformations in medieval France, pp. 33 – 60; and, most recently, Ovid's *Metamorphoses* in the School Tradition of France, pp. 48 – 82. Book One of the Vulgate Commentary has been critically edited in my PhD diss.: A Study of the "Vulgate" Commentary; and my translation of Book One is published in *The Vulgate Commentary on Ovid's Metamorphoses: Translation of Book One*. The Vulgate Commentary is also treated in Zeeman, In the Schoolroom, pp. 1 – 18.

2 The manuscripts of the Vulgate Commentary are fully listed in Coulson and Roy, *Incipitarium Ovidianum*, no. 421.

3 See especially McKinley, *Reading the Ovidian Heroine*. Regius's commentary was first printed in Venice in 1493; a critical edition of Books 1–IV is newly published in Benedetti, ed., *Raffaele Regio, In Ovidii Metamorphosin Enarrationes I (Libri 1–IV).*

4 For Arnulf, see especially Ghisalberti, Arnolfo d'Orléans, pp. 157 – 234; and Coulson and Nawotka, The rediscovery of Arnulf of Orléans's glosses, pp. 267 – 99. David Gura of the University of Notre Dame is currently working on an edition of the entire commentary. See his Ph.D. diss., A critical edition and study of Arnulf of Orléans's philological commentary to Ovid's Metamorphoses.

5 For Bersuire's influential *Ovidius moralizatus*, see Moss, *Ovid in Renaissance France*; and Coulson and Roy, *Incipitarium Ovidianum*, no. 2 (with full bibliography).

of the Vulgate Commentary to the explication of Ovid's epic, namely how our commentator references prose authors and poets nearly contemporaneous with him either to bolster an interpretation of the Ovidian passage under discussion, or to elucidate Ovidian influence on Latin writers of the Middle Ages. But before examining this specific question more closely, allow me to say a few words about the Vulgate Commentary in order to place the commentary more clearly within its cultural and intellectual milieu.

The Vulgate Commentary on the *Metamorphoses* was a product of the rich school tradition on classical poetry which developed in the Loire valley of France during the twelfth and thirteenth centuries, most notably at the cathedral school at Orléans.[6] Internal references in the commentary itself, as well as the date of the earliest exemplars of the commentary, place the date of its composition to the mid-thirteenth century.[7] The importance of the Vulgate Commentary for our understanding of the school tradition on Ovid was signalled as early as 1920 by Luigi Castiglioni.[8] While Castiglioni knew of five witnesses to the text, my own research on the commentary, conducted over the last twenty years, has uncovered twenty-two extant witnesses.

One reason why the Vulgate Commentary may be considered so central to the commentary tradition on Ovid is that it synthesizes the medieval scholarship on the poet from the eleventh century down to the mid-thirteenth century. Many of its glosses and longer comments are drawn directly from earlier commentaries of German origin, now preserved in manuscripts at the Bayerische Staatsbibliothek in Munich.[9] Additionally, the sources for many of the allegories are drawn from the earlier work of Arnulf, who was an influential master at Orléans in the later twelfth century. Grammatical glosses adapted directly from the commentary of William, another master at Orléans nearly contemporary with Arnulf, are also in evidence in the commentary.[10] No single unified method characterizes the approach adopted by the Vulgate commentator. Rather he employs a highly varied

6 The bibliography on Orléans as a center for the transmission of the classics is extensive. See, in particular, Engelbrecht, *Carmina Pieridum*, pp. 209–26; and *idem*, Fulco, Arnulf, and William, pp. 52–73.

7 The manuscripts of the Vulgate Commentary are fully discussed and described in Coulson, MSS. of the *'Vulgate'* commentary, pp. 118–29.

8 See his Spogli riccardiani, pp. 162–66.

9 The principal manuscripts are Munich, Bayerische Staatsbibliothek, clm 4610, clm 14482, and clm 14809. Robin Wahlsten Böckerman, a doctoral candidate at the University of Sweden, is editing these texts.

10 William's commentary is most fully discussed in Engelbrecht, *Filologie de Dertiende eeuw.*

and eclectic approach to the text, dealing with the rudimentary questions of grammar and syntax; more advanced problems related to myth and science; and, as one might expect, allegorical interpretations which explicate various transformations morally or historically.

But it is undoubtedly for its literary sensitivity that the Vulgate Commentary may claim a unique position among the hundreds of commentaries and random glosses transmitted in the margins of the manuscripts of the *Metamorphoses* during the thirteenth and fourteenth centuries. When we speak of the "literary interests" of the commentary, we include such aspects as character delineation in the epic, structural techniques used by Ovid to provide cohesion and unity to the fifteen books, the manner in which Ovid develops specific scenes within the *Metamorphoses*, and comments which discuss Ovid's choice of diction, as well as references to parallel passages in classical Latin authors which serve to elucidate aspects of Ovid's Latinity. To give the flavour of the range of comments, allow me to examine three glosses which reflect the generally high level of interpretation of the commentator.

At *Met.* 7.658 Cephalus and his troops are delayed from returning to Athens by the wind that blows from the east. This ill-fated wind gives Ovid the opportunity to indulge in one of his favourite narrative devices, that of the banquet, in which numerous tales are narrated in a sort of imbedded narrative. At the beginning of Book Eight, some 300 lines later, Ovid picks up the forward momentum of the narrative with the words, *Dant placidi cursum redeuntibus austri/ Aeacidis Cephaloque* (The steady south wind speeds the return of the allies of Aeacus and Cephalus). The Vulgate Commentary specifically underlines this reference and elucidates for the reader how the line serves as an important structural marker. For our second example, let us turn to the story of Niobe, the queen who loses her seven daughters and sons because of her overweening pride. At *Met.* 6.275, Ovid uses the adjective *resupina*, which here means haughty, to describe her character. A quick perusal of the Oxford Latin Dictionary reveals that this meaning of the word is rather unusual. The Vulgate Commentary not only properly stresses the unusual meaning but points the reader back to the beginning of the story at *Met.* 6.166, where Niobe had also been described as *spectabilis*. Finally, there are many scenes in the *Metamorphoses* where Ovid seeks to develop thematic connections with earlier ones, as is the case with the moral dilemma faced by Procne in killing her son Itys and that of Althaea, who ultimately kills her son Meleager to revenge the death of her brothers. The Vulgate Commentary examines in some detail stylistic similarities between the two scenes.

Let us now turn our attention to the specific topic under investigation in this article, namely the manner in which the Vulgate commentator discusses Ovidian influence on the poetry and prose of the twelfth-century renaissance. The comments which follow are based on a relatively full transcription of the commentary which I was able to make while a graduate student at the University of Toronto. The Vulgate commentator alludes specifically to the following twelfth- and thirteenth-century writers in the commentary: Alain of Lille; Bartholomaeus Anglicus; Bernard Silvester; the author of the *Pamphilus de amore*; the *Tobias* of Matthew of Vendôme; Everhard of Béthune's *Graecismus*; Alexander de Villa-Dei's *Doctrinale*; and Walter of Châtillon's *Alexandreis.*

At its most basic level, the Vulgate commentator draws upon authors nearly contemporaneous with him to bolster an allegorical interpretation or to explicate more fully the exact meaning of a word. So, for example, early in Book 1 of the *Metamorphoses*, Ovid describes the assault made upon the heavens by the Giants (*Met.* 1.151–162), who are defeated and whose blood causes a regeneration. Our commentator allegorically explains this as an illustration of the fact that wickedness breeds wickedness. He strengthens his initial interpretation, which is indebted to the *Allegoriae* of Arnulf of Orléans, by quoting from the *Tobias* of Matthew of Vendôme and the *Pamphilus de amore:*

> Intelligendum est allegorice. Per illos qui nati sunt de sanguine, quod de reprobis nascuntur reprobi. Arbor enim mala bonum fructum non potest facere. Vnde magister Matheus: Arbor fronde patet. Quod confirmat Pamphilus dicens: Premonstrat signis patrem natura frequenter. (Vat. lat. 1598, f. 3r)
>
> [We should understand this transformation allegorically. From those who were born from the blood understand that wicked people are born from wicked people. For an evil tree cannot bear good fruit. (Matt. 7:18) Whence Master Matthew says, "A tree is made manifest by its leaf." (*Tobias* 1164–65) And Pamphilus confirms this saying, "Nature frequently shows its progenitor in clear signs." (*Pamphilus de amore* 351)]

This is a technique which is employed relatively extensively in the allegories, as may be seen by a further example, this time taken from the allegory to *Met.* 5.328:

> Mutationes iste morales sunt. Dii enim fugati a gigantibus, id est hominibus ge, id est terram, amantibus, in Egiptum fugientes, in diuersas se mutauerunt figuras, quod ideo fingitur quod Egiptii potius fuerunt ydolatre quam alii et diversas diversorum deorum statuas habuerunt, et hoc fecit Ieroboan, filius Salomonis, qui preerat Egipto, quem in regem super se constituerunt decem tribus, recedentes a Roboan per dissensionem; qui Ieroboan, timuns ne secundum iussum Moysi ascenderent in Ierusalem et relinquerent eum, fecit duos vitulos aureos et constituit utrique favum suum, alterum in Dan, aliud in Betel, et dixit ad populum: Ecce dii tui, Israel, qui eduxerunt te de terra Egypti et hoc est quod magister Galterus commemorat in Alexandro dicens: ne tamen infamet gentem et genus, ydola regum, sordes Samarie, fraterni numina regni preterit. (Vat. lat. 1598, f. 50r)
>
> [The transformations are moral. For the gods, put to flight by the giants, that is men, lovers of ge, that is to say, the earth, fled to Egypt and transformed themselves into various

shapes. And one imagines this because the Egyptians were greater idolaters than others and had various statues of diverse gods. And Jeroboam did this, the son of Salomon, who was commander of Egypt. The ten tribes placed him as king over them, withdrawing from Roboam through dissent. And this Jeroboam, fearing lest according to the order of Moses, they should go up into Jerusalem and abandon him, made two golden calves and set up an offering for each, one in Dan, the other in Bethel, and he said to the people: Behold your gods, Israel, who led you from the land of Egypt, and teacher Walter recalls this event in his *Alexandreis* saying: lest such infamy should stain the tribe, he passes over the royal idols, gods of the realm, Samaria's disgrace. (*Alex.* 4.242–44)]

Similarly, our commentator at times has recourse to contemporary poets or prose authors to explain an unusual word or concept in Ovid's text. For example, at the beginning of the *Metamorphoses*, Ovid describes the world before creation as a sort of pea soup without clear distinctions:

Ante mare et terras et quod tegit omnia caelum,
unus erat toto naturae uultus in orbe,
quem dixere Chaos, rudis indigesta moles. (*Met.* 1.5–7)

[Before the ocean, or earth, or heaven, Nature was all alike, shapeless, chaos so called, all ruddy and lumpy matter, inert and confused.]

Here the Vulgate commentator has recourse to Bartholomaeus Anglicus's elaborate description of chaos from Book 8 of the *De rerum proprietatibus*, as well as to the more purely literary description of Chaos found in Book 1 of the *Cosmographia* of Bernard Silvester (at 1.20–23):

Met. 1.5 *ante mare:* Postquam expediuit se auctor a propositione et inuocatione, accedit ad narrationem, sed ad maiorem euidenciam subsequencium quedam uideamus que ualde sunt utilia. Bartholomeus autem dicit ita de yle: "Yle est globus informis, sine loco, sine tempore, sine quantitate, sine qualitate, inter aliquam et nullam substantiam." (Vat. lat. 1598, f. 1r)

[*Before the sea* etc.: After the author extricated himself from the proposition and the invocation, he proceeds to the narration. But for a better understanding of what follows, let us investigate certain things that are most useful. Bartholomew, moreover, states the following about Hyle: "Hyle is an unformed globe, without place, without time, without quantity, without quality, between some and no substance."]

The reference to the *Cosmographia* of Bernard Silvester follows:

1.5 *Vnus erat uultus:* una uoluntas nature. Natura esse uolebat unum tantummodo secundum confusionem, modo non habet unum uelle natura quia aliud uult in bruto, aliud in homine. Vel uoluntas quod uult magister Bernardus dicens:
Turbida temperiem, formam rudis, hispida cultum

optat et a ueteri cupiens exire tumultu
artifices numero<s> et musica uincla requirit. (Vat. lat. 1598, f. 1r)

[*There was a single visage:* one inclination of nature. Nature wanted to be one only in regards to confusion; now nature cannot desire to be one thing, since nature is different in man and in animals. Or he uses inclination (*uoluntas*) with the same meaning as Bernard Silvester who says,
Full of movement, she desires moderation; ill-formed, she desires shape;
Shaggy, she seeks adornment; desiring to escape the ancient tumult,
She seeks artful proportions and the bonds of harmony. (*Cosmographia*, 1.20 – 22)]

Similarly, at *Met.* 4.58, the storyteller alludes to Queen Semiramis, whose background the Vulgate commentator explicates with reference to Walter of Châtillon's similar appeal at *Alex.* 6.33:

Semiramis regina fuit a qua denominata fuit regio Semiramia, que fecit fieri Babilonem, quod etiam innuit magister Galterus: ergo Semiramis postquam Mauortius heros finibus egressus etc. (Vat. lat. 1598, f. 35r)

[Semiramis was a queen and the Babylonian region is named after her; and she constructed Babylon, and teacher Walter alludes to this: Therefore after the warlike hero had quit the lands of Semiramis etc. (*Alex.* 6.33)]

At other times, the Vulgate Commentary alludes to contemporary poets to elucidate more specifically an Ovidian influence on a particular scene. So, for example, early in Book 1, Ovid provides an elaborate description of the Milky Way along which the gods proceed to the council convened by Jupiter:

Est via sublimis, caelo manifesta sereno;
lactea nomen habet, candore notabilis ipso,
hac iter est superis ad magni tecta tonantis
regalemque domum, dextra laevaque deorum
atria nobilium valvis celebrantur apertis;
plebs habitat diversa locis. (*Met.* 1.168 – 73)

[Easily seen when the night skies are clear, the Milky Way shines white. Along this road the gods move toward the palace of the Thunderer and his royal halls, and on the right and on the left, the dwelling of the gods are open and guests come thronging, while the lesser gods live in meaner sections.]

The Vulgate commentator remarks on the similarity of presentation to be found in the *Anticlaudianus* of Alain of Lille, 5.403 – 407.

Est uia: magister Alanus de eodem idem sonat:
Set quoniam totus scintillat in igne beato
hic locus et flamme nutu blanditur amico

censetur polus empireus cui flamma benignis
ignibus arridet aulamque nitoribus ornat. (Vat. lat. 1598, f. 3r)

[So Alain of Lille uses similar language in his description:
But since this whole place sparkles with blessed fire and is soothed by the friendly swaying flame, it is identified with the empyrean pole on which the glow of a kindly fire smiles and sets off the court with its lustre.]

Ovid, moreover, favoured elaborate and detailed descriptions of the *locus amoenus*, such as we find in the elaborate descriptions of Diana's bathing pool (at *Met.* 3.155) and of the fountain where Callisto's pregnancy is revealed (at *Met.* 2.455). Walter of Châtillon appears to incorporate elements of these descriptions into his own ecphrasis in the *Alexandreis*, specifically at Book 2. 306 ff:

Met. 2.455 similis erat isti rivo de quo dicit magister Galterus: verisque latens sub ueste iocatur riuulus et longo rigat interiora meatu garrulus et strepitu facit obsurdescere montes. (Vat. lat. 1598, f. 18r)

[Like that river which master Walter describes: The playful brook hides in spring's cloak and waters the inner reaches with its long wandering and its noisy chatter makes the mountains deaf. (*Alex.* 2. 314)]

Met. 3.161 *fons:* simile dicit magister Galterus: fons cadit a leva quam cespite gramen adumbrat purpureo. (Vat. lat. 1598, f. 26r)

[*a fountain:* like what master Walter says: a fountain falls from the left where purple grasses cast their shadow. *(Alex.* 2.313–14)]

At other places, the verbal appropriation of his Ovidian model by a twelfth-century poet is explicitly stressed. The comments which the Vulgate Commentary provides to *Met.* 1.108 and 1.128 are perhaps the most telling examples of this trait which I have found. At *Met.* 1.108–11, Ovid describes the Golden Age as one in which men are free from toil and enjoy an endless spring, since the year has not yet been divided into the four seasons:

Ver erat aeternum, placidique tepentibus auris
mulcebant Zephyri natos sine semine flores;
mox etiam fruges tellus inarata ferebat,
nec renouatus ager grauidis canebat aristis.

[Spring was forever and a west wind blew softly across the flowers that sprung unplanted; the untilled earth brought forth grain and the fields unfallowed grew white with rich wheat.]

The Vulgate Commentary alludes to the similarity in presentation to be found in Walter of Châtillon's *Alexandreis* at 2.317–318:

Mulcebant: magister Galterus, uolens sapere uim uerborum Ouidii, locum delectabilem in quo Darius ad suos loquitur describit dicens:
Hic mater Cybele, Zephirum cui, Flora, maritans pullulat, et uallem fecundat gratia fontis.
(Vat. lat. 1598, f. 2r)

[Walter, wishing the might of Ovid's description to be redolent, describes similarly a delectable place in which Darius addresses his troops:
Here wedding the Zephyr to you Flora, Mother Cybele bears fruit and the graciousness of the fountain enriches the valley.]

At *Met.* 1.128 – 32, Ovid concludes his enumeration of the four ages of man by describing the age of iron as follows:

de duro est ultima ferro.
protinus inrupit venae peioris in aevum
omne nefas, fugere pudor verumque fidesque.
in quorum subiere locum fraudesque dolique
insidiaeque et vis et amor sceleratus habendi.

[The Iron Age followed, whose base vein let loose all evil: modesty and truth and righteousness fled earth, and in their place came trickery and slyness, plotting, swindling, violence, and the damned desire of having.]

The Vulgate comments thus on Walter's apparent appropriation of these lines:

Protinus: memoriter tenens uerba Ouidii magister Galterus similia hiis dixit:
Pululat humanum genus et polluta propago;
decedit virtus, vicium succedit, adherent
coniugio illicito, pietas rectumque recedunt. (Vat. lat. 1598, f. 2v)

[Holding in memory the words of Ovid, Walter develops his scene similarly:
The human race increased and the human race was polluted. Virtue withdrew and vice took its place. People clung to illicit unions. Dutifulness and righteousness departed. (*Alex.* 4.195 – 97)]

And lastly, in Book 1, line 150 of the *Metamorphoses*, Ovid comments that the goddess Astraea is the last to flee the earth with the advent of the Iron Age. The Vulgate comments on the use made of this motif by Walter at *Alex.* 1.176 – 77:

Reliquit ad hanc reuocandam inuitare uidetur magister Aristotiles Alexandrum:
Et per te reuocetur ab alto ultima que superum terras Astrea reliquit.
(Vat. lat. 1598, f. 3r)

[Aristotle appears to invite Alexander to recall this very goddess when he says:
let Astraea, last of the gods to leave the earth, be recalled from heaven by you.]

In other places the Vulgate commentator astutely references places, particularly in the poetry of Walter of Châtillon, where one can deduce a nearly exact Ovidian borrowing. For example, at *Alex.* 2.354, Darius encourages his men with phrases seemingly borrowed directly from *Met.* 3. 548:

> *Patrium retinete decus:* Simile dicit Darius militibus suis: ergo agite, o proceres, patrium reuocate uigorem (Vat. lat. 1598, f. 31v).
>
> [Hold onto your native honor: so Darius addressing his troops: therefore act, men, recall your native strength. (*Alex.* 2.354)]

And most tellingly, in his description of the Hermaphrodite's transformation (at *Met.* 4. 379), Ovid employs a unique phrase, *neutrumque utrumque*, which is picked up by Walter of Châtillon:

> 4.379 *neutrumque utrumque:* quia partim uir, partim femina; ab his uerbis habuit occasionem describendi crepusculum magister Galterus dicens: tempus erat dubiam cogens pallescere lucem , cui neque lux neque nox imposuit nomen, utrumque et neutrum tenui discrimine etc. (Vat. lat. 1598, f. 39v)
>
> [*neither and both:* because he was part man, part woman; Walter takes up this phrase in describing dusk: it was a time that compelled the faint light to grow pale, to which neither day nor night could give its name, both and neither with slight distinction etc. (*Alex.* 3.463–65)]

In several places, the Vulgate commentator cleverly deduces Walter's use of an Ovidian coinage, that is to say a word used in a slightly unusual or unique way. At *Met.* 9.75–76, Hercules addresses the shape-shifter Achelous, who has transformed himself into a snake to avoid capture by Hercules. Hercules rejoins: "What's in store for you, do you think, you who are no proper snake at all, no weapons of your own, just skulking in a shape you've begged and borrowed" (*Quid fore te credis, falsum qui versus in anguem/ arma aliena moves, quem forma precaria celat?*) The adjective *precaria* here has a somewhat unusual meaning of borrowed, and the Vulgate commentator produces the following gloss:

> Simile in *Alexandreide:* inter honoratos fulgere precaria uestis. (Vat. lat. 1598, f. 90r)
>
> [Likewise in the *Alexandreis:* Shone in their borrowed clothes (*Alex.* 5.469][11]

The commentator here emphasizes the exact similarity in meaning between the use of the adjective in Ovid and Walter.

---

11 The phrase reoccurs at *Alex.* 6.346.

And finally, there are places where the more straightforward allusion in Ovid is reinterpreted allegorically through what might be termed a "Christian filter" in order to recontextualize it. For example, in Book 10 of the *Metamorphoses*, Orpheus goes to the Underworld in search of Eurydice. Ovid here gives us what one might call an American Express tour of the region. At *Met.* 10.34 – 35, Orpheus in his song to Pluto and Persephone says: "You are Lords over these regions in which human kind is doomed to linger for the longest time" (*vosque/ humani generis longissima regna tenetis*). The commentator draws the connection between Ovid's use of the adjective *longissima* to describe Hades and Walter's description of Hell in Book 10 of the *Alexandreis*, where Satan calls an assembly of the gods of the Underworld:

> *Met.*10.35 *longissima:* ad penas respicit infernales que nunquam defficiunt. De quibus dicit magister Galterus in *Alexandreyde:* O supplicium miserabile semper/et numquam moritur quem torquet carcer Averni. (Vat. lat. 1598, f. 100v)
>
> [*longest:* refers to the punishments of the underworld which never cease. Walter speaks of this in the *Alexandreis:* the man who is tortured in Hell is always and never dying. (*Alex.* 10.119 – 20)]

A similar connection is drawn by the Vulgate commentator between Walter's Christian conception of Hell at *Alex.* 5.140 and Ovid's more pagan description from Book 3 of the *Metamorphoses*:

> *Met.* 4.442 *omnes animas:* ideo magister Galterus: et umbriferi domus insatiabilis Orci (*sic*). (Vat. lat. 1598, f. 40r)
>
> [*all souls:* hence teacher Walter: and the insatiable home of shady Hades. (*Alex.* 5.141)]

The Vulgate Commentary therefore demonstrates a detailed and sustained exposition of the influence of Ovidian style and word usage on the poets and prose authors of the twelfth-century renaissance, ranging from the relatively more jejune comments on the manner in which Alain of Lille, Walter of Châtillon or Bernard Silvester may have appropriated an Ovidian mannerism or style to the far more sophisticated treatment of how an Ovidian neologism or coinage has been taken up by the medieval poet (as was the case with the meaning of *precaria* discussed above). The Vulgate commentary thus has much to reveal concerning the circulation and manner in which these celebrated French poets were read and studied by contemporary critics.

Bibliography

Alain of Lille, *Anticlaudianus : texte critique avec une introduction et des tables*, ed. R. Bossuat, Textes philosophiques du môyen âge 1, Paris: Librairie Philosophique J. Vrin, 1955.

Bartholomaeus Anglicus, *De rerum proprietatibus*, Frankfurt, 1601.

Arnulf of Orléans, *Allegoriae*, in Fausto Ghisalberti, Arnolfo d'Orléans: un cultore di Ovidio nel secolo XII, *Memorie del Reale Istituto Lombardo di Scienze e lettere* 24 (1932): 157–234.

M. Benedetti, ed., *Raffaele Regio, In Ovidii Metamorphosin Enarrationes I (Libri 1–IV)*, Florence: SISMEL, Edizioni del Galluzzo, 2008.

L. Castiglioni, Spogli riccardiani, *Bollettino di filologia classica* 27, 1920, pp. 162–66.

F.T. Coulson, Ovid's transformations in medieval France (ca. 1100–ca. 1350), in Alison Keith and Stephen Rupp, eds., *Metamorphosis: The Changing Face of Ovid in Medieval and Early Modern Europe*, Toronto: Centre for Reformation and Renaissance Studies, 2007, pp. 33–60.

F.T. Coulson, Ovid's *Metamorphoses* in the School Tradition of France, 1180–1400: Texts, Manuscript Traditions, Manuscript Settings, in James Clark, Frank T. Coulson, and Kathryn McKinley, eds., *Ovid in the Middle Ages*, Cambridge, Cambridge University Press, 2011, pp. 48–82.

F.T. Coulson, MSS. of the "Vulgate" commentary on Ovid's *Metamorphoses:* A checklist, *Scriptorium* 39, 1985, pp. 118–29.

F.T. Coulson, A study of the "Vulgate" Commentary on Ovid's *Metamorphoses* and a critical edition of the glosses to Book One, Ph.D. diss., Univ. of Toronto, 1982.

F.T. Coulson, *The Vulgate Commentary on Ovid's Metamorphoses: Book One*, Kalamazoo: Western Michigan University Medieval Institute, 2015.

F.T. Coulson, and K. Nawotka, The rediscovery of Arnulf of Orléans' glosses to Ovid's creation myth, *Classica et Mediaevalia* 44, 1993, pp. 267–99.

F.T. Coulson and B. Roy, *Incipitarium Ovidianum: A Finding Guide for Texts related to the Study of Ovid in the Middle Ages and Renaissance*, Turnhout: Brepols, 2000.

W. Engelbrecht, *Carmina Pieridum multo vigilata labore/exponi, nulla certius urbe reor.* Orléans and the reception of Ovid in the aetas Ovidiana in school commentaries, *Mittellateinishes Jahrbuch* 41, 2006, pp. 209–26.

W. Engelbrecht, *Filologie de Dertiende eeuw: De Bursarii super Ovidios van Magister Willem van Orléans (fl. 1200 AD)*, Olomouc: Nákladatelství Univerzity Palackého, 2003.

W. Engelbrecht, Fulco, Arnulf, and William: Twelfth-century Views on Ovid in Orléans, *The Journal of Medieval Latin* 18, 2008, pp. 52–73.

F. Ghisalberti, Arnolfo d'Orléans: un cultore di Ovidio nel secolo XII, *Memorie del Reale Istituto Lombardo di Scienze e lettere* 24, 1932, pp. 157–234.

D. Gura, A critical edition and study of Arnulf of Orléans's philological commentary to Ovid's Metamorphoses, Ph.D. diss., The Ohio State University, 2010.

Matthew of Vendôme, *Tobias*, in Franco Munari, ed., *Mathei Vindocinensis Opera*, vol. 2, *Piramus et Tisbe, Milo, Epistule, Tobias*, Rome: Edizioni di storia e letteratura, 1982.

K.L. McKinley, *Reading the Ovidian Heroine: Metamorphoses Commentaries 1100–1618*, Leiden: Brill, 2001.

A. Moss, *Ovid in Renaissance France*, London: Warburg Institute, 1982.

Ovid, *P. Ovidii Nasonis Metamorphoses*, ed. Richard J. Tarrant, Oxford: Oxford University Press, 2004.
Pamphilus de amore, in F. Bertini, ed., *Commedie latine del XII e XIII secolo*, vol. 3, Genoa: Pubblicazioni dell'Istituto di filologia classica dell'Università di Genova, 1980.
Bernardus Silvestris, *Cosmographia*, ed. P. Dronke, Textus minores 53, Leiden: E.J. Brill, 1978.
The Vulgate Commentary (Vatican City, Biblioteca Apostolica Vaticana, Vat. Lat. 1598).
Walter of Châtillon, *Galteri de Castellione Alexandreis*, ed. M. Colker, *Thesaurus mundi* 17, Padua: Antenore, 1978.
N. Zeeman, In the Schoolroom with the *Vulgate Commentary* on *Metamorphoses* I, *New Medieval Literatures* 11, 2009, pp. 1–18.

Michael McVaugh

On the Individuality of the Medieval Translator

In a way perhaps it is paradoxical to talk about the individuality of medieval translators. After all (though I doubt that they thought of it in this way), many medieval translators took it as a goal in effect to suppress individuality – to reproduce exactly and word-for-word in Latin, say, the text that they had before them in Arabic or Greek. But of course we are all aware that the process of translation is more open-ended than they supposed, made up of a virtually infinite number of translatorial decisions that are all shaped by the translator's knowledge and experience and motives – marks therefore of his individuality. We can identify some of these decisions when we compare a Latin text with its Arabic original and recognize that a translator chose to translate *ʿaql* with *intellectus* rather than *ratio*,[1] for example; but that does not often help us explain those choices in terms of the translator's background or personality unless we already know something about that from independent sources. Fortunately, one of the tasks that students of medieval translations have long set themselves is to accumulate such sources – biographical or autobiographical statements, say, or archival records, or references in colophons. We all appreciate the thoughtfulness of Gerard of Cremona's *socii* in preparing for us a précis of his life and works.[2]

This paper is meant to offer reflections on a way in which studies of the translations in themselves can sometimes offer the historian what we might call a "cross-bearing" on the individuality of the translator. This has been brought home to me by my own very haphazard studies of translations, which have come to suggest to me – not by any means a particularly original insight – that certain kinds of comparative studies have the potential to be particularly rewarding. They have this potential because (if I can put it this way), rather than showing us a translator picking one translation out of an indeterminate number of conceivable possibilities, they show us a translator picking one translation in preference to another sometimes very concrete possible choice.

One such kind of comparative study might be termed diachronic, and is possible when a translator is known to have produced a number of translations over a period of time. Does his language change over that period, and if so, do the changes have anything to tell us about his motives or the development of his ca-

1 Akasoy, Die lateinischen Übersetzungen, pp. 689–701, at p. 697.
2 Burnett, Coherence, pp. 249–88; the biography is edited at pp. 273–87.

reer or his social context – his individuality, that is? I have been working for some years with both Gerrit Bos and Joseph Shatzmiller on the history of a short text by Galen, *De inequali temperie* or *De malicia complexionis diverse.* We have been looking collectively at its passage from an Arabic translation (by Ḥunayn b. Isḥāq), to a Latin translation of the Arabic (by Gerard of Cremona), to a Hebrew translation from the Latin (by David Caslari, about 1310), preparing editions of the three, each one in the light of the version from which it had been translated.[3]

I happened to notice idly in the course of this study that Gerard always translated the Arabic phrase *wa dhālika anna* as "et illud ideo quoniam;" but subsequently I became aware that in his Aristotelian translations he always translated the same phrase in a second, quite different way, as "quod est quia" or "quod est quoniam." I then examined as many of his translations as I could, comparing them with the Arabic original whenever it was accessible, and came to the realization – to greatly oversimplify – that they can mostly be divided into those two forms, with a few translations seeming to occupy a transitional position.[4]

This kind of data obviously raises a number of possibilities for further study. Might they reveal a change in Gerard's own translational technique over the course of his career, and if so, in which direction? From more literal ("et illud ideo quoniam") to freer and less literal ("quod est quia")? Might the two classes hint at divisions within a Gerardian atelier? I am not trying to draw big conclusions from this very simple discovery; I want simply to suggest that this kind of comparative study may allow us to talk about a translator's personal development over time.

I might offer a slightly different instance of "diachronic" translation for consideration. The late-thirteenth-century physician Arnau de Vilanova grew up in Valencia in the years immediately following its reconquest by Jaume I; he knew Arabic and translated several medical works from that language. In two cases the Arabic original has already been edited, and Arnau's Latin translations of those works have been, or will be, edited for inclusion in the ongoing edition of his Latin medical works. One is a treatise on medicinal simples by Abū al-Ṣalt, the other a treatise on cardiac medicines by Avicenna (Ibn Sīnā). We have no independent evidence as to when in his career these were produced, whether in Arnau's early years in Valencia and in university study down to perhaps 1280;

3 This study has recently been published as Bos, McVaugh, and Shatzmiller, *Transmitting a Text.*

4 McVaugh, Towards a Stylistic Grouping, pp. 99–112.

or in the 1280s, when he was physician to the kings of Aragon (and when a third translation of his from Arabic was produced); or in the 1290s, when he was teaching at Montpellier. We do not even know which of the two was the earlier. But if we compare the translatorial techniques involved in the two we can see striking differences that may not entirely be the results of the very different character of the underlying works. In Abū al-Ṣalt's text, a simple compilation of medicines and their properties, Arnau abbreviates a certain amount of detail and makes surprisingly crude mistakes at the beginning, though he corrects them as the work progresses. Avicenna's *De viribus cordis* is a much more sophisticated piece of writing; half of it is a philosophical treatise explaining the nature of medicinal action in terms of Galenic-Aristotelian physiology, yet Arnau's translation is clear and exact from the outset – it has no problem with technical terminology or argument, it is not tied to word-for-word presentation, and it knows (for example) that when Avicenna refers to the *Kitāb al-sūfīsṭīqān* [*sic*] the Aristotelian *De sophisticis elenchis* is in question.

Set against the quite remarkable amount of independent detail that we have about Arnau's intellectual and professional development, a comparison of these two translations should certainly have a story to tell us – but *what* story? Is the Abū al-Ṣalt translation an early Valencian production by Arnau (Abū al-Ṣalt himself was Valencian, dying only a century before Arnau grew up there), when Arnau was still developing linguistically and intellectually; and is the Avicenna translation instead a mature work produced in 1280, when Arnau was in touch with an arabizing community in Barcelona? Or, as Danielle Jacquart has suggested, might the weaknesses in the former translation be the consequence of a rushed production while Arnau was at Montpellier in the 1290s, a quick and dirty set of notes made not for publication but for his own personal use, in which his Arabic started coming back to him as he proceeded?[5] Again, I am not trying to resolve the question; I am trying, instead, to suggest the kinds of information that may possibly be extracted from a comparative study of one translator's work, and that deserve to be part of our exploration of his individuality.

So one kind of comparative study looks at a single translator over time – comparing him with himself, so to speak. A second kind – "synchronic," I suppose – compares two translators at work independently on the same base text, to see what decisions each translator made at specific points in the same text, and, perhaps (in the light of other knowledge), why. I found myself engaged with this kind of material when Gerrit Bos suggested that I edit the Latin texts of Maimo-

5 See Abū al-Ṣalt and Arnaldus de Villanova, *Translatio libri Albuzale*, pp. 430–1, 446.

nides's minor medical works (none had ever been printed), as an aid to reconstructing the Arabic and Hebrew versions that he was preparing.

As it happens, between 1295 and 1305 there were two Latin translators of Maimonides, both physicians, active simultaneously and essentially independently: Armengaud Blaise (Arnau de Vilanova's nephew), working at Montpellier, and Giovanni da Capua, working at Rome. Both left introductions to their translations, and we know quite a bit about each of them: Armengaud was a Montpellier academic who translated from Arabic and (with help) from Hebrew, and dedicated one translation to Pope Clement V; Giovanni was a converted Jew who translated from Hebrew (and perhaps Arabic), and dedicated some of his work to Pope Boniface VIII. But their translations, looked at side by side, tell us even more about them. Both have left us a Latin translation of Maimonides's little treatise *On Asthma*, and both seem to have translated it from the original Arabic; but their versions, while generally intelligent and accurate, are nevertheless distinctively different, and our ability to compare them on points of detail seems to reveal features of individual personality. Thus, for example, Armengaud's translation is consistently ten percent longer than Giovanni's. Not only does it use more words to say the same thing in Latin, its vocabulary is ten percent larger. Whether unconsciously or deliberately, Armengaud was writing a fuller, more polished – or at least more fluid, certainly more articulate – Latin. Giovanni fell back on the vague word "res" thirty-eight times in translating Maimonides's text; Armengaud used it only a dozen times. Armengaud seems more often to use words reflecting the ambience of an academic medical faculty, and he repeatedly adds language that in effect gives a deliberate and original clinical gloss to Maimonides's statements – in effect giving his views on why, or where, the drugs that Maimonides recommends have an effect.[6]

I find it particularly interesting that Armengaud's exposure from childhood to the Latin liturgy of the Christian church seems to have left traces in his phraseology. Translating Maimonides's dramatic account of the palace reaction to a Moroccan ruler's death from an overdose of theriac, Armengaud put a mixture of New Testament passages into his Jewish author's mouth: "Factus est repente [Acts 2:2] in palatio clamor magnus [Matthew 25.6];" Giovanni, the Jewish convert, simply rendered the passage as "audita est vox fletus."[7] Were these scriptural echoes conscious or unconscious?

A similar echo in Armengaud's version of another Maimonidean work, *On Hemorrhoids*, is almost certainly conscious. In *De emorroidibus* 7.3, Maimonides

6 Maimonides, *On Asthma* (vol. 2), pp. xxxviii – xlii.
7 Ibid., p. xlv.

is describing the technique of suffumigation, which involves placing an earthenware pot (Ar. *qaṣriyya*) with a hole in its base upside down over smoking coals for the patient to sit on and absorb the smoke rectally. Armengaud chose to use the unusual word *perapsidis* to translate "pot" – not, one would have thought, an obvious term to select, certainly not one in common domestic or medical use. It must surely derive from Matthew 23:25–26, where Christ commands the Pharisee: "munda prius quod intus est calicis et paropsidis [Gr. *paropsidos*], ut fiat id quod deforis est mundum," i.e., "cleanse first what is within the cup and platter, that the outside of them may be clean also." It may be far-fetched to wonder if the "cleansing" image could have evoked the word to Armengaud as he was describing this particular medical treatment, but it certainly confirms that, for whatever reason, scriptural terminology came easily into his mind while he was in the process of translating. Giovanni da Capua translated the same work, and he prosaically rendered *qaṣriyya* as "vas."[8] At the moment, we happen to know much more from other sources about Armengaud's life than we do about Giovanni's, which is why it is easier to perceive Armengaud's translation as a reflection of that life, and, in turn, to deepen our understanding of his personality, shaped by the academic and ecclesiastical world he had undoubtedly grown up in. But the contrast with Giovanni's usage sharpens our perception.

I have opted to devote the rest of this paper to one particular – and peculiar – case, one which may further help illuminate the Catalan world of Arnau and Armengaud as a theatre of transmission. Unlike my earlier examples, it concerns medieval translation *from* Latin rather than *into* Latin, but I have chosen to treat it – and in some detail – because it illustrates my argument well, and shows what students of transmission can hope to find when they look carefully at its individual agents. I have been interested in the tradition of learned Latin surgery for some time. It is a tradition running from the mid-thirteenth to the mid-fourteenth centuries, taking its authority from the writings of Galen and Albucasis (Abū al-Qāsim Khalaf al-Zahrāwī), and many works in this series were soon translated into vernacular European languages by its admirers.[9] Among the earliest of these translations were two, made into Catalan, of the *Chirurgia* of Teodorico Borgognoni, which was perhaps the most influential of the new scientific surgeries. The original was completed about 1265. The two translations were both finished by 1315, and each survives today in a single manuscript. One (now in Paris) is a translation composed by the experienced surgeon Guillem Corretger,

8 Maimonides, *On Hemorrhoids*, p. xlii.
9 McVaugh, *Rational Surgery*.

ca. 1300;[10] the other (now in Graz) is a text by a young Montpellier medical master named Bernat de Berriac and is dated in its colophon to the year 1311.[11] I have recently examined these two versions in order to understand how they are related to Teodorico's work, and whether they are related to each other.[12]

A first detailed comparison of the two texts established that Bernat's version was largely based on Guillem's original – that, in fact, he made very few changes to Guillem's translation of the first three books of Teodorico's work (merely breaking them down into smaller units), but prepared an entirely new version of the fourth book. It would be consistent with what we know from other sources about their circumstances and careers to suppose that Bernat encountered Guillem and his translation when they were both studying at Montpellier (ca. 1303), that Bernat began to copy out the translation there, but was unable to complete his work before leaving to begin medical practice in Catalonia and then launch a career as physician to the kings of Mallorca, and that he was eventually able to finish it by making his own translation of the last book directly from a copy of Teodorico's Latin.

So that fourth book of Teodorico's *Chirurgia* again puts before us two simultaneous and independent translations of a single text. They are both, I might add, quite faithful, rendering the Latin nearly word-by-word (and Catalan word order and syntax is not all that different from medieval Latin); and, as we go through it in the two versions, we see – as we would expect – our translators making different choices about terminology:

Teodorico	**Guillem (Paris MS)**	**Bernat (Graz MS)**
fractura	trencament	fractura
esophagum	ysofagun	asbrer
reuma	cadarn	reuma
tempora	temples	gautes

One might have guessed that the surgeon would be motivated to choose vernacular terms, and the academic physician to choose closer Latinate cognates, espe-

10 The text is in MS Paris, Bibliothèque nationale, espagnol 212, ff. 1ra – 89vb. A digitized copy is accessible at http://gallica.bnf.fr/ark:/12148/btv1b8438665 s.

11 On the manuscript and its history, see Kern, *Die Handschriften*, pp. 199 – 200, available (at the date of this writing, 11 Feb. 2016), at http://www.manuscripta-mediaevalia.de/hs/katalogseiten/HSK0514a_b199_jpg.htm. The manuscript itself is accessible in digitized form at http://143.50.26.142/digbib/handschriften/Ms.0200 – 0399/Ms.0342/index.html. See also L. Cifuentes, Vernacularization, 133 – 34.

12 A full account of this investigation is given in McVaugh, Academic Medicine.

cially since an authorial preface to Guillem's translation explains that he decided to prepare it for the benefit of his practitioner-colleagues who were unable to read Latin. In fact, however, in this limited sample the contrast is not clear-cut. As these examples show, both of our translators seem to have opted, now for one kind of term, now for the other. I am sure that when the two versions are carefully edited and systematically compared as a whole, we will be able to say more about the motives behind their individual choices, but that lies in the future.

But this is not all that the two versions have for us to think about. A still closer examination of the Graz manuscript revealed that the many alterations and marginal emendations made there to the text of the first three books (there are none at all in the fourth) were apparently Bernat's own later corrections to the text that he had originally taken over essentially unaltered from Guillem's version – and, in fact, that this very manuscript is probably Bernat's own personal copy, so that the physical codex itself, and not merely the translation, dates from 1311. We might extend my earlier explanatory model and imagine that, after Bernat finished translating the fourth book from the Latin Teodorico he had managed to acquire, he went back through the first three books comparing this Latin with Guillem's Catalan version of the text, emending the Catalan where he felt it desirable. If this hypothesis is correct – and I really believe that it is – the Graz manuscript shows us a translator's actual draft; it shows him in the very act of making specific decisions about meaning and specific editorial guesses about the underlying text, and the sum of these decisions gives us a window on the individual, on his motives and his abilities.

Bernat did not leave a preface for posterity, like Guillem, to let us know about his personal motives for translation, but I think we can still speculate intelligently about them. In my view, he is doing this for himself, because he has perceived the utility of a vernacular text to his practice. I will not here classify his changes in detail, but I will say that while some of them are corrections of errors that he perceives, others are to bring the text into harmony with the contemporary academic medical discourse he had been trained in. For example: in many medieval Latin recipes the Greek word *ana* is regularly used in the sense of "of each, of every one alike," in specifying a quantity applicable to every ingredient, and Teodorico's Latin routinely used this word in recipes. When Guillem came to this word, he must have felt that his surgical colleagues might not understand it, and he always used one of a number of different Catalan phrases to translate it: "de totes aquestes," "de cascun," "aytant de la .i. com del altro," "eugualment," and so forth. But in each of these passages Bernat restored the original technical word *ana* that he was used to. Unlike Guillem, he was not thinking of a wider audience; if I am right, this is his own original copy, and we have no evidence that it was ever copied and so passed into wider circulation.

Still, it was read carefully at least once more, as some unusual marginal notes prove. Throughout Book I of the *Surgery*, the Catalan chapter titles in Bernat's translation of Teodorico have been rendered into Hebrew. We can imagine, I think, that when Bernat died in the plague of 1349 his book passed into the hands of a Mallorcan Jew, a physician, who could read the Catalan text but found that adding Hebrew chapter titles was a useful tool for finding his way around in it. (By the sixteenth century the volume had gone north, to Austria, but the Hebrew is in a Sephardic rather than Ashkenazi hand.) We know enough about Jewish physicians in the Mallorcan kingdom (and their possession of medical and surgical books) for this to seem wholly plausible[13] – and it implies, I should point out, that by this time written Hebrew had become not merely a religious language, but a scientific vernacular, so that a Jew in a Catalan-speaking land actually found it easier to read medical Hebrew than Catalan.

In addition to the chapter titles, the manuscript contains one much longer marginal note in Hebrew.[14] A few scholars have recognized that this note contains a series of three recipes, attached to Teodorico's discussion of consolidative medicines (medicines used to restore flesh and skin); what they have not been able to recognize is that in fact the recipes are a Hebrew translation from another source, the *Chirurgia parva* of Lanfranc of Milan. This is another work in the tradition of learned Latin surgery, a work originally composed about 1290.[15] Our Mallorcan Jew expanded Bernat's Catalan description of such medicines by adding three more relevant remedies from the new European surgery – but in Hebrew! And he thus appears to have provided me with a further case that bears on the point of this paper: might even these fragmentary translations, translations of two different works – the Catalan Teodorico and the Latin Lanfranc – apparently by the same man, have something to tell us about him as an individual?

To answer this is made unusually difficult by the linguistic character of the time and place in which he lived; this leads to a puzzle that has to be resolved before we can definitely assess the abilities of our Mallorcan Jew. The Catalan world of the first half of the fourteenth century was one in which Latin, Catalan, and Hebrew could all be languages of science, and transmission between them was commonplace. Lanfranc's *Chirurgia parva* was itself translated into Catalan

13 A useful introduction is Contreras Mas, *Los médicos judíos*.
14 MS Graz, Universitätsbibliothek, 342, f. 25v.
15 Lanfranc of Milan, *Chirurgia Parva*, where the passage in question is on f. 206.

by the Montpellier physician Guillem Selva, in 1329. Might not our Jewish physician have been translating the recipes, not from the original Latin, but from the Catalan that we know he could read? It is by no means impossible, but I think it unlikely for a number of reasons. One piece of evidence I find particularly telling: at the end of one of the recipes in the translated passage, Lanfranc's Latin version recommends reserving the product for future use, "servetur usui." Selva's translation recognizes this perfectly well by rendering it "stoial" [= "stoia-lo," today "estojar"].[16] But the Hebrew translation of the passage in the Graz manuscript says instead "offer it," not "reserve it." A translator from Latin could have confused "servire" and "servare," a confusion that would be easy for someone with a relatively limited knowledge of Latin; but it is not obvious what Catalan word he could have confused with "stoia" that would have meant "offer" – and this therefore makes it seem more likely that he was translating from a Latin than from a Catalan original. Does this Hebrew selection from the *Chirurgia parva*, then, in conjunction with the marginal chapter titles, reveal our Mallorcan physician to us as a man able to draw on medical sources in both Latin and Catalan, reasonably comfortable in both languages, but happiest when converting them into Hebrew?

Perhaps... but perhaps not, because we have not quite come to the end of the puzzle. Over a century ago, Moritz Steinschneider pointed to the existence of a medieval Hebrew version of the *Chirurgia parva*, of which he identified four copies – in Munich, Oxford, Cambridge, and Lyons;[17] since then, three more (Turin, Vatican, Oxford) have turned up. Immediately one might wonder whether the Hebrew is based on Lanfranc's original or on Selva's Catalan version, for this would further complicate the study of our Mallorcan Jew. Might he have worked, not from a Latin original, not even from a Catalan translation, but from a Hebrew translation instead, jotting down Lanfranc's recipes in the form he found them there? In fact, the preliminary answer is one for which our glances at the transmission process have not prepared us: the Hebrew characters in which the text is written seem to contain the text itself in Romance, not in the Hebrew language,[18] and the version was thus meant for a reader who was literate in Hebrew and

16 ".... Apres prenla entre les mans untades ab oly rosat he longament sia meloxat he stoial et si es estiu mesclay ab la rasina aygua et la meytat de cera" (Lanfranc of Milan and Guillem Selva, *Suma de cirurgia*, f. 49 – 49v). The Latin original reads ".... Postea sumatur inter manus inunctas oleo ros. et diu manibus malaxetur, et servetur usui, sed in aestate admisce cum resina quantum est medietas cerae" (Lanfranc of Milan, *Chirurgia Parva*, f. 206).

17 Steinschneider, *Die Hebraeischen Übersetzungen*, pp. 807 – 8.

18 Joseph Shatzmiller, personal communication (based on his inspection of the Hebrew translation as contained in MS Munich, chm 280, ff. 257 – 62).

could speak but not read the Romance tongue – and such a reader clearly does not correspond to the annotator of the Graz manuscript of the Catalan Teodorico.

Thus, in the end our little scrap of Hebrew text has exposed to us an astonishingly complex pattern of transmissions of a single text – of pathways for its communication to recipients who might read Hebrew or Latin script or both, who might understand spoken but not written Catalan, who could comprehend more than one written language (Latin or Catalan or Hebrew) but used them for different purposes. Our concern in the studies in this volume has of course been with agency, and agency's forms and motives share correspondingly in this complexity – indeed, the author of that scrap was arguably both agent and recipient. Our best understanding for the moment must be that he was a trilingual physician, able to work directly with Latinate medicine but most at home in Hebrew; yet it is still possible that he was a physician who valued the new European surgery but needed to seek out Hebrew or Romance texts to be most comfortable with it. Either way, our puzzle has something to teach us about the interrelationships of linguistic cultures in fourteenth-century Mallorca, and Catalunya more broadly. But of course the puzzle already illustrates, *as* a puzzle, how a series of translations from the same person – even tiny fragments! – has the potential, when taken in conjunction with other information, to reveal some distinctive aspects of that person's individuality.

Bibliography

Abū al-Ṣalt and Arnaldus de Villanova, *Arnaldi de Villanova Opera Medica Omnia, XVII: Translatio libri Albuzale de medicinis simplicibus*, ed. J. Martinez Gazquez, Ana Labarta, Michael McVaugh, Danielle Jacquart, and Lluís Cifuentes, Barcelona: University of Barcelona, 2005.

A.A. Akasoy, Die lateinischen Übersetzungen der Risāla fī 'l-ʿaql al-Kindīs, in Maria Cândida Pacheco and José F. Meirinhos, eds., *Intellect et imagination dans la philosophie médiévale*, 3 vols., SIEPM Rencontres de philosophie médiévale 11, Turnhout: Brepols, 2006, vol. 1, pp. 689–701.

G. Bos, M. McVaugh, and J. Shatzmiller, *Transmitting a Text Through Three Languages: The Future History of Galen's 'Peri Anomalou Dyskrasias,'* Transactions of the American Philosophical Society, vol. 104, part 5, Philadelphia: American Philosophical Society Press, 2015.

C. Burnett, The Coherence of the Arabic-Latin Translation Program in Toledo in the Twelfth Century, *Science in Context* 14, 2001, pp. 249–88.

Ll. Cifuentes, Vernacularization as an Intellectual and Social Bridge, *Early Science and Medicine* 4, 1999, pp. 127–48.

A. Contreras Mas, *Los médicos judíos en la Mallorca bajomedieval, siglos XIV–XV*, Palma de Mallorca: Font, 1997.

A. Kern, *Die Handschriften der Universitätsbibliothek Graz*, vol. 1, Leipzig: Otto Harrassowitz, 1942.

Lanfranc of Milan, *Chirurgia Parva,* in *Ars chirurgica Guidonis Cauliaci*, Venice 1546, ff. 201r–207r.

Lanfranc of Milan and Guillem Selva, *Suma de cirurgia*, MS Madrid, Biblioteca nacional 10162, ff. 1–54.

Maimonides, *On Asthma*, vol. 2, ed. Gerrit Bos and Michael R. McVaugh, Provo: Brigham Young University Press, 2008.

Maimonides, *On Hemorrhoids*, ed. Gerrit Bos and Michael R. McVaugh, Provo: Brigham Young University Press, 2012.

M.R. McVaugh, Academic Medicine and the Vernacularization of Medieval Surgery: The Case of Bernat de Berriac, in *El saber i les llengües vernacles a l'època de Llull i Eximenis*, ed. Anna Alberni, Lola Badia, Lluís Cifuentes, and Alexander Fidora, Barcelona: Abadia de Montserrat, 2012, pp. 257–81.

M.R. McVaugh, *The Rational Surgery of the Middle Ages*, Florence: Edizioni del Galluzzo, 2006.

M.R. McVaugh, Towards a Stylistic Grouping of the Translations of Gerard of Cremona, *Mediaeval Studies* 71, 2009, pp. 99–112.

M. Steinschneider, *Die Hebraeischen Übersetzungen des Mittelalters und die Juden als Dolmetscher*, Berlin: Kommissionsverlag des Bibliographischen Bureau, 1893 [reprint: Graz: Akademische Druk- U. Verlagsanstalt, 1956].

Teodorico Borgognoni and Bernat de Berriac, *Libre de sirorgia qui es dit tadorich*, MS Graz, Universitätsbibliothek, 342, ff. 1–282.

Teodorico Borgognoni and Guillem Corretger, *Libre lo qual compila frare thederich*, MS Paris, Bibliothèque nationale, espagnol 212, ff. 1–93v.

Raphaela Veit

Charles I of Anjou as Initiator of the *Liber Continens* Translation: Patronage Between Foreign Affairs and Medical Interest

> However, after long courses of years, that most Christian ruler, Charles, King of Jerusalem and Sicily, was moved by the fame of this book and was attracted by the desire for such an easily accessible utility. And he decided, not only for his own advantage but also for the benefit of all those who honour Christ, to unite his martial energies with his care for knowledge in his efforts towards a translation of this book. Thus in thankfulness for the kingdoms and mighty esteem that he had received through God's will in glorious battles he did gratefully present to the people a gift that honoured their founder and sustainer. After he had brought the aforesaid book from the King of Tunis by means of a ceremonial delegation into his possession, he found a reliable man, who had mastered the Arabic and the Latin language and consigned to him the book, whose value to us lay hidden in the darkness of its Arabic language. And he illuminated this book, with the light of a Latin translation.[1]

This quotation originates from the introduction to the translation of Rhazes's *Kitāb al-Ḥāwī fī al-Ṭibb* (*The Comprehensive Book on Medicine*), which was commissioned by Charles of Anjou, the ruler – at least nominally – of Anjou, Provence, Jerusalem, Achaia and the Kingdom of Sicily.[2] The work, in Latin *Liber Continens*, is a medical encyclopaedia of one of the most significant scholars of the Islamic world, whose reputation as a doctor was particularly widespread. This was Abū Bakr Muḥammad b. Zakariyyāʾ al-Rāzī, born in 864 in Ray, not far from Teheran, where he died in 925.[3]

The prologue clearly portrays Charles as the initiator of the translation. In addition to this we also learn the extent of his efforts to find the Arabic manuscript, by means of a ceremonial delegation to the so-called King of Tunisia. This

---

1 Fischer and Weisser, Das Vorwort, p. 270, § 20: *Sed post longua [sic] annorum curricula christianissimus Dominus Karolus, Iherusalem et Sicilie Rex, eiusdem libri fama conmotus et ardore tam captabilis utilitatis illectus, non sibi solum sed christicolis omnibus prodesse desiderans, bellicis curis liberalium studiorum sollicitudinem ad ipsius libri translationem miscere delegit, ut in partem recognitionis regnorum et titulorum ingentium, que divinitus sibi bellorum fuerant acquisita triumphis, referret aliquid populis in eius honorem qui populorum est conditor et seruator. Predicto itaque libro a rege Tunisii per sollempnes nuntios conquisito virum fidelem adhibuit lingue tam arabice quam latine peritum, et in libro ipso, in quo sub arabice lingue tenebris tanta nobis occultabatur utilitas, mandavit et fecit lucernam latine translationis accendi.*

2 Ibid., pp. 211 – 18; Dunbabin, *Charles I of Anjou*, pp. 222 – 4.

3 Ullmann, *Die Medizin im Islam*, pp. 128 – 36.

passage *[a rege Tunisii per sollempnes nuntios conquisito]* is usually interpreted as a reference to the delegation of the ruler of Anjou. However, it was also translated as "emissaries of the emir of Tunis;"[4] the wording permits both translations. Furthermore we read that Charles also located a translator for the manuscript, referred to as "a reliable man, who had mastered the Arabic and the Latin language."

The information given in this introduction is picked up again in the illuminations of the oldest manuscript of the Latin version, which can now be found in Paris (BNF lat. 6912, written 1279–1282). The first in the series of miniatures shows the three emissaries with the so-called "King of Tunis," who is shown handing the manuscript over to them. The clothing of the emissaries gives the impression that they are members of Charles's court and not that of the Emir. In the next picture the emissaries, now arrived in Naples, are shown handing the book over to Charles, who is then pictured handing it over to the translator in the upper half of the third picture. The translator is then portrayed as absorbed in the work of translating the text.[5]

No information is given about the translator in the introduction. However, at the end of the manuscript in the *explicit* the following reference is made:

> Thus ends the translation of the book El Havy on medicine [...], which the commissary of the most excellent King Charles ordered to be written [...] by the hand of the master Faragius, a Jew, who was his devoted translator and the son of a Salernic master from Agrigento [...] on Monday, the thirteenth of February, in the seventh Indiction, near to Naples [...].[6]

This allows us to draw conclusions regarding, firstly, the name of the translator: a Jewish master named Farag of Agrigento, his father originating from Salerno. The *explicit* tells us secondly the date that the translation was completed, which was the thirteenth of February 1279.[7] The papers from the chancellery of Charles show that the translation was begun on the sixth of February 1278.[8] Given the breadth of Rhazes's *Kitāb al-Ḥāwī* and the fact that it was translated within a

4 This is Dunbabin's translation on the back cover of *Charles I of Anjou.*

5 Avril, Gosset, and Rabel, *Manuscrits enluminés*, plate CXIII (= Ms. Paris, BNF lat. 6912, vol. 1, f. 1v).

6 Ibid., p. 159 (no. 186) (= Ms. Paris, BNF lat. 6912, vol. 1, f. 289v): *Explicit translatio libri El Havy in medicina* [...] *facta de mandato excellentissimi regis Karoli* [...] *per manus magistri Faragii judei filii magistri salerni de agrigento devoti interpretis eius* [...] *die lune XIII° februarii VII indictione apud Neapolim* [...].

7 Ibid.

8 Fischer and Weisser, Das Vorwort, p. 240, §20; Cohn, Jüdische Übersetzer, p. 250 [p. 54 of the reprint].

year, it seems unlikely that Farag was working alone. This assumption is supported by our knowledge of the numerous writers and copiers employed by the Neapolitan Court[9] – why would Farag have been the only one of these who knew Arabic? We know of at least one other Jewish translator present at Charles's court who worked on Arabic texts: this was Moses of Palermo, who is recorded as having translated a pseudo-Hippocratic work for Charles.[10] Farag has been identified with the Jew Faraj b. Sālim but we know very little about his biography.[11]

In the quote given above, Faraj described himself as *judei filii magistri salerni de agrigento.* Maybe his mention of Salerno is an allusion to his father's contact with the Medical School in the city. We also find, in an earlier decree of Charles to the Jewish community of Palermo and Garbo the name of a certain Ferragut who examined – and passed – the future rabbi of the community.[12] Whether the Ferragut referred to here and the translator Faraj b. Sālim are one and the same must remain an open question. However what can be said with certainty is that following his translation of Rhazes's work, Faraj continued to undertake translation commissions from the Neapolitan Court. His *De expositionibus vocabulorum seu sinonimorum simplicis medicine* probably has to be seen as a part or as an appendix to *Liber Continens,*[13] and another work was his translation of the tractate *Taqwīm al-abdān fī tadbīr al-insān* (*Tacuin agritudinum / Dispositio corporum de constitutione hominis*), by the eleventh-century Baghdad physician Abū

---

9 Fischer and Weisser, Das Vorwort, p. 239, §20.

10 On Moses of Palermo, who is said to have translated Ps.-Hippocrates's *Liber de curationibus infirmitatum equorum,* see Cohn, Jüdische Übersetzer, pp. 259–60 [pp. 63–4 of the reprint]; Roth, Jewish Intellectual Life, p. 320; Fischer and Weisser, Das Vorwort, p. 240, §20; Fischer, Moses of Palermo, pp. 278–81; and Arieti, Mosè da Palermo, pp. 55–63. For an intellectual classification of Moses see Zonta, The Jewish Mediation, p. 92; as well as Freudenthal, Arabic and Latin Cultures, p. 86.

11 For the identification of Farag or Farasius or Farragut / Ferragut or Farresche, see Steinschneider, Donnolo, pp. 296–336. Already in 1935 the Jewish scholar Willy Cohn from Breslau gathered more or less all the available information about this figure, and I am grateful to Gad Freudenthal for the reference to this valuable article, which is rarely mentioned in the research literature on the subject (Cohn, Jüdische Übersetzer, pp. 248–58 [pp. 52–62 of the reprint]). On Willy Cohn, see Freudenthal, Arabic and Latin Cultures, p. 85, n. 34; on Faraj b. Sālim, see also Roth, Jewish Intellectual Life, p. 320.

12 Cohn, Jüdische Übersetzer, p. 250 [p. 54 of the reprint].

13 On this discussion, see Ibid., p. 62 and n. 34 [reprint].

ʿAlī Yaḥyā b. ʿĪsā b. ʿAlī b. Jazla, an astronomically-influenced work on medicine.[14]

Going back to Charles of Anjou and the *Liber Continens:* in the list of repeated payments to clerks, copyists, translators, and illuminators given in Charles's decrees, we find a reference to a payment made to Faraj for his translation. We are informed about the progress of the work in the same context; Faraj was obviously working on the fifth and last book of the translation in September 1278. The original Arabic manuscript appears to have consisted of five volumes, which were held by Charles's treasurer. It seems that only the volume currently undergoing translation was permitted to leave his custody.[15] In 1279 the treasurer was given instructions to entrust to Faraj two large trunks which appear to have served as the repository of the translation and were to be held in the Castel d'Uovo:

> Order the treasurer to give "*Fareche le juif deux bons coffres grans et larges, lesquex nous voulons que vous pregniez en nostre chastiau de Salvateur en mer, liquel est etc.; et nous envoiez (par ledit Fa)-reche les cinc livres encians d'arabis qui vindrent de Tunes, et les cinc premier livres en arabic des vint .... chastiau de l'Euf a vos, quant nous en partimes derreannement...*"*). Donné a la Tour de Capes, le XXV jour de fevrier de la VII*^{e} *indicion.*[16]

Not only this; the foreword to the Latin translation gives us further information about how Charles continued after the completion of the translation work:

> He [Charles] submitted this [translation] for the examination, review, and correction of no lesser authorities than the professors of the art of healing, his court physicians and others at the universities of Naples and Salerno. For this process he awarded them not a short, but an appropriate period of time. On the appointed day all these authorities reported, openly and each for himself on the translation. Their unanimous vote and their concordant opinion awarded the translator and the compiler [of the work] the highest praise. [...][17]

---

14 Ibid., pp. 254, 257 [pp. 58, 61 of the reprint]; and Ullmann, *Die Medizin im Islam*, p. 160, with more information on this convert from Islam to Christianity. See also Veit, Transferts scientifiques.

15 Cohn, Jüdische Übersetzer, p. 55 [reprint].

16 R. Filangieri, ed., *I Registri della Cancelleria Angioina*, vol. 21, pp. 1278–9, 208 (no. 43: Reg. 34, f. 48). The orthography and punctuation marks in the citation follow the text given by Filangieri.

17 Fischer and Weisser, Das Vorwort, p. 227, § 21: *Hanc nichilominus artis medicine professorum familiarium medicorum suorum ac aliorum Neapoli et Salerni regentium examinationi provisioni et correctioni supposuit, et eis ad id non summarium, set ordinarium tempus indulsit. Qui omnes et singuli de translatione huiusmodi statuto ad hoc die relationem in publico facientes una voce et una omnium concordi sententia dignissimos laudum preconiis extulerunt interpretem cum auctore.*

A critic might question how much philological expertise the specialists named here used for their evaluation of the Latin translation of this comprehensive Arabic work.[18] Charles, in any case, is bound to have been pleased with the unanimous praise that his translation initiative was awarded. However, his patronage did not end with the completion of the translation and its review by an expert panel.

He spared neither money nor effort to have the translation copied as a richly decorated manuscript. This was the codex mentioned earlier, which is now located in Paris (BNF lat. 6912). The work for this manuscript was directed by the French doctor and librarian of Charles of Anjou, Jean de Nesle. Details from the archives of the Anjou chancellery reveal that eight copyists and two miniature painters participated in the work. The informative illuminations at the beginning of the manuscript were carried out by the monk Giovanni of Monte Cassino, one of the most famous miniature painters of his time, whilst most of the illuminated letters can be attributed to Minardus Theutonicus. In his function as ruler of Achaia, Charles also commissioned a further copy, via the Chancellor of Morea. We have a significant amount of information about this copy from the chancellor decrees as well. It is most likely that this copy became the manuscripts found today in the Vatican library under the reference numbers Vat. lat. 2398 and 2399. They appear to have been copied following the example of the Parisian manuscript BNF lat. 6912; not only were at least some of the same copyists employed for the process, but corrections found on the margin of the Parisian manuscript have also been worked into the text. Again, part of the illuminations of the Vatican copy can be attributed to Giovanni of Monte Cassino.[19] For the later translations from Faraj, parchment was supplied by Charles; furthermore the king ordered seven copies of the *Taqwīm*.[20] But what was Charles's motivation for his activity as an initiator of Latin translations of Arabic medical texts?

The introduction to the Latin translation clearly reveals Charles as the person responsible for commissioning the work. Traditionally this introduction is seen as the work of Faraj, the translator.[21] But the academic, elevated tone of the introduction is of a different level than that of the language found in the actual translation. The reason may be that the authorship of the prologue was also

18 See Fischer and Weisser (*op. cit.*), who do question this point.
19 On this complicated context, see the detailed discussions of Avril, Gosset, and Rabel, *Manuscrits enluminés*, pp. 158–9; and Cohn, Jüdische Übersetzer, pp. 252–3, 257–8 [pp. 56–7, 61–2 of the reprint].
20 Cohn, Jüdische Übersetzer, p. 256 [p. 60 of the reprint].
21 See ibid., p. 251 [p. 55 and n. 14 of the reprint].

located in the courtly academic, medically trained circle around Charles.[22] However, this difference could also be explained by the difficulty of producing a translation from the Arabic into a more or less literal but also elegant Latin. Whatever the reason, it is clear that the phrases used in the prologue are intended to unambiguously emphasize Charles's role as patron of sciences. We find a portrayal of Charles as a ruler who loves sciences and translates his love into action, even when this is bound up with significant cost and effort. The monetary aspect is already clear in the luxurious decoration of the manuscript that Charles had commissioned (Ms. Paris, BNF lat. 6912).

Historically, the presentation of Charles of Anjou is dominated by the image of an unscrupulous Machiavellian, who exercised extensive influence over the fate of the Mediterranean world in the second half of the thirteenth century.[23] This evaluation often ignores the long tradition of knowledge transfer on which Sicily and lower Italy look back, a tradition in which Charles is seamlessly integrated. Under the Norman rulers and the Hohenstaufen ruler Emperor Frederick II some texts were translated, not only from Greek, but also from Arabic. We should remember Frederick's interest in falconry, which he mainly pursued through Arabic texts (e. g., extracts of Avicenna's biology), the Latin translations of which were then incorporated into his famous book on falconry.[24] Frederick's son Manfred, who was at least temporarily recognised as his successor, continued the tradition of his father, commissioning and paying for the Latin translation of scientific texts, including Arabic works.[25] As such, we should not be surprised by Charles's commitment; in Sicily and lower Italy he found a translation tradition and the necessary infrastructure for this, which only required adopting and expanding.

It is also worth noting that certain elements of the population that Charles found in the Kingdom of Sicily were helpful for compiling Arabic translations;

22 Fischer and Weisser, Das Vorwort, p. 211.

23 Dunbabin, *Charles I of Anjou*, pp. vii–viii.

24 S. Georges, *Das zweite Falkenbuch*; Kaiser Friedrichs II, *Das Falkenbuch Friedrichs II*; Fried, *Kaiser Friedrich II.* On the intellectual milieu in medieval Sicily (with respect to Latin translations) see also Veit, Transferts scientifiques.

25 King Manfred of Sicily ordered the Latin translation of one praxis-related book that was regarded equally highly in both east and west: the *Taqwīm al-Ṣiḥḥa* (The Almanac of Health) by the Iraqi Ibn Buṭlān (d. 1038). A number of illustrated manuscripts survive from the Middle Ages. See: Biedermann, *Medicina magica*; and Ibn Buṭlān, *Tacuinum sanitatis in medicina* (vol. 13). The fact that this "do-it-yourself" healthcare as well as the *Taqwīm* of Ibn Jazla (supra, n. 14) were translated into German as early as the sixteenth century reflects the continued popularity and influence of Arabic praxis on early-modern Europe (Ullmann, *Die Medizin im Islam*, pp. 157–8).

not only Jewish, but also some Muslim religious communities were permitted to practise their religion freely, although Charles's interest in non-Christian religious groups remained limited. Such communities were tolerated because of the practical benefits that their presence offered.[26]

When we attempt to reconstruct Charles's choice of this text in particular for translation and copying, we are obliged to consider the possibility of his fundamental interest in medicine. It is known that Charles attempted to build up Naples as a city for medical education.[27] The University of Naples was founded by Frederick II in 1224 and the medical statutes of Paris adopted for the medical faculty. However, in 1231 medical teaching was suspended, making the support of the ruler welcome.[28] In the introduction to the translation a clear reference is given to the fact that the review of the *Liber Continens* ordered by Charles was not only carried out by the scholars of the medical school of Salerno – which was transformed into a medical faculty in 1280 – but also from those in Naples.[29]

Charles's interest in medicine does not necessarily explain why he chose Rhazes's bulky and difficult work, the *Kitāb al-Ḥāwī*, for translation. The work was not overly successful in the Islamic world and was not destined for greater popularity in the West. The book is little more than a medical scrapbook; it contains a wide range of quotations from Greek authorities and their editors in Arabic, sometimes with a commentary on the subject from Rhazes himself. Obviously this wide range of quotations was found in Rhazes's possession upon his death and they were arranged by his students into a medical handbook. The clarity and organisation of this handbook leaves much to be desired – a criticism made not only by Arabic scholars such as al-Majūsī but also later by scholars in the Latin West.[30] In the Latin tradition, a new structuring of the content of the encyclopaedia was eventually undertaken, in the context of the printing of the text in early-modern Europe.[31]

Why choose this work, which was neither easy nor user-friendly? In the ninth and tenth centuries, three medical encyclopaedias originated in the Islamic cultural area: the *Almansor* (*Kitāb al-Manṣūrī*) of Rhazes, the *Liber regalis* (*Kitāb Kāmil al-Ṣināʿa al-Ṭibbiyya / Kitāb al-Malakī*) of al-Majūsī, and the *Canon* (*al-*

26 Dunbabin, *Charles I of Anjou*, p. 154.
27 Fischer and Weisser, Das Vorwort, pp. 240 – 1, §21.
28 Ibid.; for more information see Monti, *Storia dell'Università di Napoli*.
29 Veit, *Das Buch der Fieber*, pp. 225 – 6.
30 Fischer and Weisser, 212 – 14; Cohn, 24 – 248 / 51 – 52; Ullmann, 130 and 143 – 44.
31 al-Rāzī and Hieronymus Surianus, *Continens Rasis ordinatus et correctus per clarissimum artium et medicine doctorem magistrum Hieronymum Surianum*. On the problems which arise from the new structuring of the content, see Veit, Greek roots, Arab authoring, Latin overlay, p. 368.

Qānūn fī al-Ṭibb) of Avicenna. All of these works had already been translated into Latin by the time that Charles came to rule,[32] leading to the question: did Charles want to make a comparable work available to the Latin medical scholars?

Charles's choice of a significant work from the field of Islamic medicine as the subject for the Latin translation which would make him famous may also have been coincidence. Whatever the reason, there was obviously no manuscript in the libraries of south Italy and Sicily that fulfilled Charles's requirements for an impressive translation. A couple of points should be noted here.

There is a thoroughly positive attitude towards science and research that the Qurʾān and Ḥadīth had instilled into the entire Muslim community. In various times and places, various centres for knowledge where these aspects were highly valued and supported did spring up, and these centres often stood in competition with one another. Baghdad, Damascus, Cairo, Kairouan, Fez, and Córdoba in Andalusia can be named as examples. All of these centres were distinguished by the patronage of a caliph or a particularly powerful local ruler for whom the support of the sciences played a role in their power struggle with the caliph. Returning to the regions that are now part of Italy and Spain, we are able to establish that – from an Islamic-historic perspective – Spain and its caliphal court of Córdoba was one of the most significant intellectual centres of the Islamic world. In contrast, Italy only experienced a short period of Muslim rule – on the island of Sicily – and Sicily never attained the rank of a centre for knowledge as Córdoba did.[33]

As a logical consequence, the amount of scientific material found in the Spanish archives and libraries which only required translation for the Latin world to make use of it must have been of an almost incomprehensible quantity. By contrast, in Sicily the contents of the libraries were somewhat more modest, and most of value was taken by Muslim refugees to North Africa when the Norman invasions drove them out of Sicily.[34] A patron in south Italy would therefore not only have to establish the infrastructure for translations, but first and foremost to obtain the Arabic originals that would be used for the work. Consequently Charles first needed to locate an Arabic manuscript of the *Kitāb al-Ḥāwī*. This

32 On the Latin translations of the *Almansor* and the *Canon* done by Gerard of Cremona in twelfth-century Toledo, see Jacquart and Micheau, *La médecine arabe et l'occident médiéval*, p. 150. On the two Latin translations of al-Majūsī's work, see Ullmann, *Die Medizin im Islam*, p. 132; and also Veit, Ibn al-Jazzār's *Viaticum*, [forthcoming].
33 Houben, *Roger II. von Sizilien*, p. 119, n. 32.
34 Ibid., pp. 105 – 6.

problem may be seen in the context of questions about Charles and his relationship to the Orient.

Charles's contacts with the region known today as Tunisia are characterised by a number of aspects. Firstly, Charles was particularly active in the thirteenth-century politics of the Mediterranean area even before he began his reign over south Italy. It was this that brought him such dubious pleasures as a temporary imprisonment under the Mamluks, for example. Later he laid great importance on achieving the title – at least nominally – of King of Jerusalem.[35] The experiences of his youth surely nurtured his interest in the Orient and consequently in Arabic scientific texts.

Following Charles's conquest of Sicily and lower Italy in 1266, Tunisia became particularly important for a number of reasons. Firstly, ever since Norman rule over the island, the ruler of Sicily had exercised a claim over Tunisia – although this claim had ceased to be applied. Secondly, there were followers of the Staufer dynasty who had fled to Tunis when Charles conquered the island. A number of families in Sicily who had lost power, influence, and wealth through the downfall of the Staufers supported the refugees in Tunis, along with advocates of their interests (who mostly served the Ḥafṣids as Christian mercenaries). Charles was concerned to bring the trouble spot under control. The seventh crusade of 1270 can be understood in this context; it was actually led by the King of France – who was later named Saint Louis, and was the elder brother of Charles – against Tunis, upon the urging of Charles. Louis and a large proportion of the army died from illness during a siege of the city, but Charles still managed to conclude a contract with the Ḥafṣid emir of Tunis. The contract ensured the ejection of the Staufer sympathisers from Tunis as well as extensive tribute payments.[36] Maybe the handover of the Rhazes manuscript can also be seen in the context of these concessions. In any case, it seems unlikely that – as suggested in the prologue – Charles sent a delegation to Tunisia purely for the purpose of buying Arabic manuscripts. Rather, it seems likely this activity took place in the wider context of Charles's diplomatic and military dealings with Tunisia. Political commitments appear to have played an extremely significant role here. As discussed above, the text in the introduction to the translation may well refer to a delegation from the Emir of Tunis – and not necessarily to a delegation sent by Charles.

Despite Charles's patronage, the *Liber Continens* found only limited success in the Latin West, which was mainly due to the awkwardness of the text, com-

35 Dunbabin, *Charles I of Anjou*, pp. 89–98.
36 Ibid., pp. 195–202.

pared to the available alternatives in the field of scholastic medicine. Another reason for the less enthusiastic reception of the work may also be that – despite the efforts of Charles to expand the medical teaching centres in his realm – the real centres for medical scholarship at that time lay further north, in Paris, Montpellier, and Bologna. The heyday of the Medical School in Salerno was definitively over.[37]

In principle, the role of the patron in supporting translations cannot be praised highly enough; a translator needs the original work, additional materials to help his translation, a workplace, work instruments such as parchment and writing instruments, and, of course, a sufficient salary or some other means to support his livelihood. The role of the patrons in south Italy and Sicily was a more difficult one than that of their colleagues, in contemporary Spain, for example, because – as mentioned earlier – it seems that the libraries in south Italy and Sicily did not contain a great quantity of Arabic scientific texts. Thus a patron in south Italy not only had to create the infrastructure for the translations, but also locate the original texts and make them available to his translators.

This description of patronage in south Italy can also be applied to the period before the ruler of Anjou. Constantine the African (d. before 1098–99) is seen as the first great translator of Arabic scientific texts into Latin – not only for Italy but in the field as a whole. He is recorded as having arrived in Salerno as a refugee from Tunis with a case full of Arabic medical texts in the second half of the eleventh century. Here he was taken in by the local archbishop Alfanus and eventually transferred to the monastery Monte Cassino, where the Abbot Desiderius awarded him every conceivable help in carrying out his translations. In this case it was the translator himself who found the Arabic originals for his work, and the patrons Alfanus and Desiderius who recognised the value of the studies and acted accordingly.[38] We also know that the Norman and Staufer rulers of Sicily brought translators to their courts. These would either bring their own originals with them, or the rulers would make significant efforts to find the original

37 Veit, *Das Buch der Fieber*, pp. 225–6.
38 On Constantine the African and his Arabic books, see Hettinger, Zur Lebensgeschichte, pp. 517–29; and Veit, Quellenkundliches, pp. 121–52. A striking example of what could happen when a book was not complete or partly lost is demonstrated by Constantine's translation of the so-called *Pantegni* (al-Majūsī's *Kitāb al-Malakī*); see Green, The re-creation of Pantegni, pp. 121–60; Cartelle and Ferreira, Le De elephancia de Constantin l'Africain, pp. 233–46; Wack, ʿAlī ibn al-ʿAbbās al-Mağūsī, pp. 161–202; Veit, Al-Mağūsī's Kitāb al-Malakī, pp. 133–68.

texts for the translator to work on.[39] Charles of Anjou fits seamlessly into the tradition of the princely patronage that his predecessors established.

However, Charles of Anjou is the last south-Italian ruler who dedicated such interest to translations of Arabic. His followers, particularly Robert the Wise, paid far more attention to supporting poetry and scholarship. Thus we find ourselves in the context of Petrarch or Boccaccio, which requires a wholly different method of enquiry.[40] More generally speaking, it is clear that Charles was standing at the end of a particular process. Around 1300 the translation phase of the Greek-Arabic sciences had drawn almost entirely to a close, although the translations from Arabic texts continued to be made long into the fifteenth century. However, these translations were often new versions of works that had already undergone translation.[41]

Bibliography

S. Arieti, Mosè da Palermo e le traduzioni dei trattati di mascalcia di Ippocrate indiano, in Nicolò Bucaria, ed., *Gli Ebrei in Sicilia dal tardoantico al medioevo*, Palermo: Flaccovio editore, 1998, pp. 55–63.

F. Avril, M.-Th. Gosset, and C. Rabel, *Manuscrits enluminés d'origine italienne, Vol. 2: XIIIe siècle*, Paris: Bibliothèque Nationale de France, 1984.

H. Biedermann, *Medicina magica : metaphysische Heilmethoden in spätantiken und mittelalterlichen Handschriften*, Graz, Austria: Akademische Druck-u. Verlagsanstalt, 1978.

E.M. Cartelle and A.I.M. Ferreira, Le De elephancia de Constantin l'Africain et ses rapports avec le Pantegni, in Charles Burnett and Danielle Jacquart, eds., *Constantine the African and ʿAlī ibn al-ʿAbbās al-Maǧūsī: The Pantegni and Related Texts*, Leiden, New York, and Cologne: E.J. Brill, 1994, pp. 233–46.

W. Cohn, Jüdische Übersetzer am Hofe Karls I. von Anjou, Königs von Sizilien (1266–1285), *Monatsschrift für Geschichte und Wissenschaft des Judentums*, Jg. 79 (NF. Jg. 43), 1935, pp. 246–260 [reprinted in: *idem*, *Juden und Staufer in Unteritalien und Sizilien. Aufsätze zur Geschichte der Juden im Mittelalter, über ihr Verhältnis zu den Stauferkaisern und den Königen von Sizilien, sowie zur allgemeinen Staufergeschichte. Eine Sammlung verstreut erschienener Schriften aus den Jahren 1919–1936*, Aalen: Scientia Verlag, 1978, pp. 50–64].

J. Dunbabin, *Charles I of Anjou: Power, Kingship and State-Making in Thirteenth-Century Europe*, London and New York: Longman, 1998.

39 Houben, *Roger II. von Sizilien*, pp. 105–6.

40 Monti, *Il Mezzogiorno d'Italia nel Medioevo*, pp. 129–41.

41 Jacquart and Micheau, *La médecine arabe et l'occident médiéval*, p. 205; Veit, Avicenna's 'Canon' in East and West, pp. 335–41.

R. Filangieri, ed., *I Registri della Cancelleria Angioina*, ricostruiti da Riccardo Filangieri con la collaborazione degli archivisti Napoletani, Naples: Presso l'Accademia, 1967.

K.-D. Fischer, Moses of Palermo, Translator from the Arabic at the Court of Charles I of Anjou, *Histoire des sciences médicales* 17, 1982, (Spec 1), pp. 278–81.

K.-D. Fischer and U. Weisser, Das Vorwort zur lateinischen Übersetzung von Rhazes' Liber continens (1282): Text, Übersetzung und Erläuterungen, *Medizinhistorisches Journal* 21, 1986, pp. 211–41.

G. Freudenthal, Arabic and Latin Cultures as Resources for the Hebrew Translation Movement: Comparative Considerations, Both Quantitative and Qualitative, in *idem*, ed., *Science in Medieval Jewish Cultures*, Cambridge and New York: Cambridge University Press, 2011, pp. 74–105.

J. Fried, *Kaiser Friedrich II. als Jäger oder ein zweites Falkenbuch Kaiser Friedrichs II*, Göttingen: Vandenhoeck & Ruprecht, 1996.

S. Georges, *Das zweite Falkenbuch Kaiser Friedrichs II. Quellen. Entstehung, Überlieferung und Rezeption des Moamin. Mit einer Edition der lateinischen Überlieferung*, Berlin: Akademie-Verlag, 2008.

M.H. Green, The re-creation of Pantegni, Practica, book VIII, in Charles Burnett and Danielle Jacquart, eds., *Constantine the African and ʿAlī ibn al-ʿAbbās al-Maǧūsī: The Pantegni and Related Texts*, Leiden, New York, and Cologne: E.J. Brill, 1994, pp. 121–60.

A. Hettinger, Zur Lebensgeschichte und zum Todesdatum des Constantinus Africanus, *Deutsches Archiv für Erforschung des Mittelalters* 46, 1990, pp. 517–29.

H. Houben, *Roger II. von Sizilien. Herrscher zwischen Orient und Okzident*, Darmstadt: Wissenschaftliche Buchgesellschaft, 1997.

Ibn Buṭlān, *Tacuinum sanitatis in medicina. Glanzlichter der Buchkunst*, comm. Franz Unterkircher, Graz: Akademische Druck- und Verlagsanstalt, 2004.

D. Jacquart and F. Micheau, *La médecine arabe et l'occident médiéval*, Paris: Maisonneuve et Larose, 1990.

Kaiser Friedrichs II, *Das Falkenbuch Friedrichs II.: Cod. Pal. Lat. 1071 der Biblioteca Apostolica Vaticana*, ed. Dorothea Walz and Carl A. Willemsen, Darmstadt: Wissenschaftliche Buchgesellschaft, 2003.

G.M. Monti, *Il Mezzogiorno d'Italia nel Medioevo*, Bari: Gius, Laterza & Figli, 1930.

G.M. Monti, *Storia dell'Università di Napoli nell'Età angioina*, Naples: R. Ricciardi, 1924.

Abū Bakr al-Rāzī and Hieronymus Surianus, *Continens Rasis ordinatus et correctus per clarissimum artium et medicine doctorem magistrum Hieronymum Surianum*, pt. 1: s.l., *ca.* 1495, pt. 2: Venice: Benalius, 1509.

C. Roth, Jewish Intellectual Life in Medieval Sicily, *Jewish Quarterly Review* 47, 1957, pp. 317–35.

M. Steinschneider, Donnolo. Pharmakologische Fragmente aus dem X. Jahrhundert: nebst Beiträgen zur Literatur des Salernitaner, hauptsächlich nach handschriftlichen hebräischen Quellen, *Virchows Archiv für pathologische Anatomie* 39/2, 1867, pp. 296–336.

M. Ullmann, *Die Medizin im Islam*, Leiden: E.J. Brill, 1970.

R. Veit, Avicenna's 'Canon' in East and West: A long history of editions, *Variants* 5, 2006, pp. 331–41.

R. Veit, *Das Buch der Fieber des Isaac Israeli und seine Bedeutung im lateinischen Westen – Ein Beitrag zur Rezeption arabischer Wissenschaft im Abendland,* Stuttgart: Steiner, 2003.

R. Veit, Greek roots, Arab authoring, Latin overlay: Reflections on the sources for Avicenna's Canon, in R. Wisnovsky, et al., eds., *Vehicles of Transmission, Translation, and Transformation in Medieval Textual Culture*, Turnhout: Brepols, 2011, pp. 353–69.

R. Veit, Al-Maǧūsī's Kitāb al-Malakī and its Latin translation ascribed to Constantine the African: The reconstruction of Pantegni, Practica, Liber III, *Arabic Sciences and Philosophy* 16, 2000, pp. 133–68.

R. Veit, Quellenkundliches zu Leben und Werk von Constantinus Africanus, *Deutsches Archiv für Erforschung des Mittelalters* 59, 2003, pp. 121–52.

R. Veit, Transferts scientifiques entre Orient et Occident: Centres et acteurs en Italie médiévale (XIe–XVe s.) dans le domaine de la medicine, in R. Abdellatif, Y. Benhima, D. König, E. Ruchaud, eds., *Acteurs des transferts culturels en Méditerranée médiévale,* Munich: Oldenbourg Verlag, 2012, pp. 147–156.

M. Wack, ʿAlī ibn al-ʿAbbās al-Maǧūsī and Constantine on Love, and the Evolution of the Practica Pantegni, in Charles Burnett and Danielle Jacquart, eds., *Constantine the African and ʿAlī ibn al-ʿAbbās al-Maǧūsī: The Pantegni and Related Texts*, Leiden, New York, and Cologne: E.J. Brill, 1994, pp. 161–202.

M. Zonta, The Jewish Mediation in the Transmission of Arabo-Islamic Science and Philosophy to the Latin Middle Ages: Historical Overview and Perspectives of Research, in Andreas Speer and Lydia Wegener, eds., *Wissen uber Grenzen: Arabisches Wissen und lateinisches Mittelalter* (*Miscellanea Mediaevalia* 33), Berlin: Walter de Gruyter, 2006, pp. 89–105.

Rosa Comes

The Transmission of Azarquiel's Magic Squares in Latin Europe

The aim of this paper is to present the current state of research being carried out under the AMER project: "The Transmission of the Mathematical Magic Squares from Medieval al-Andalus to Renaissance Europe."[1] After the late Mercè Comes's identification of Azarquiel (Abū Isḥāq al-Zarqālluh/al-Zarqālī) as the Arabic source of the magic square and the corresponding text extant in the Alfonsine *Libro de Astromagia*,[2] I have focused my work on the transmission of talismanic magic square treatises to Latin Europe.[3]

As is well known, a mathematical magic square consists of a number of cells, arranged in an equal number of rows and columns, which fulfill the following condition: they must contain a string of consecutive natural numbers, starting with the number 1, and must be arranged in such a way that the sum of each row, column, and main diagonal is the same. The order of a magic square corresponds to the number of cells on each side. Therefore, a mathematical magic square of order n is a square with n cells on each side, thus n^2 cells altogether, with order 3 corresponding to the smallest possible square.[4] Finally, a magic

The research leading to this paper was carried out at the Warburg Institute of London, with the financial support of the AGAUR (Agència de Gestió d'Ajuts Universitaris i de Recerca), Generalitat de Catalunya (Spain), Research Aid 2009 – BE-003742 (second semester 2010). I would like very much to thank Professor Charles Burnett for his helpful suggestions and assistance during my stay at the Warburg Institute.

1 This research is now financed by the project: FFI2014-55537-C3-3-P, IP. R. Comes, "LA CIRCULACIÓN DE LOS MANUSCRITOS CIENTÍFICOS ÁRABES Y LATINOS EN LA CORONA DE ARAGÓN (SS. XII–XV) Y SU TRANSMISION A EUROPA", 2014–2017, MICIIN, Spain. Bink Hallum (Warburg Institute and University College London) has agreed to continue the work of the recently deceased Mercè Comes in editing the Arabic manuscripts of the texts with commentaries. Aurelie Gribomont, who is studying the only Greek manuscript, will provide its edition. As for the study of the Old Castilian and, in particular, the Latin manuscripts, I am continuing the research and editing work, together with Cristian Tolsa and Angel Martin, doctorate fellows at the University of Barcelona.

2 Ms. *Reg. Lat.* 1283ª, ed. Alfonso D'Agostino, Naples, 1992. The title, *Libro de Astromagia*, was proposed by Avilés, Two Astromagical Manuscripts, pp. 14–23.

3 For a comprehensive introduction to the subject, see Pingree, The Diffusion of Arabic Magical Texts, pp. 57–102.

4 This is because a square of order 2 is impossible; in order to comply with the main requirement that each row, column and diagonal must add up to a constant figure, all the numbers

fig. 1 Dürer's Melancholia I & Subirachs's façade of the Passion, Barcelona, Basilica of La Sagrada Familia.

square can be rotated in any direction (up, down, left, or right) or presented as a mirror image, and is always considered the same.

An example of a magic square of order 4 is shown in the right upper corner of the engraving *Melancholia I* signed by the German Renaissance master Albrecht Dürer, in which the date of the engraving, 1514, coincides with the numbers in the two central cells of the last row. A modern example of the *symbolic usage* of magic squares – although it does not comply with the rules of a magic square – is to be found in the façade of the Passion, of the basilica of La Sagrada Familia in Barcelona: its rows, columns and main diagonals add up to 33 – Christ's age at the time of his death.[5]

Two main types of treatise deal with mathematical magic squares: (1) mathematical treatises, showing a mathematical description of their construction;

would have to be equal; a square of order 2 would fail to meet the indispensable condition that all the numbers be consecutive and, thus, different. On the other hand, the square of order 3 is not representative, since out of the eight possible appearances three are the result of rotation and four of inversion of the same square, regardless of the method used to construct it.

5 The sculptor is J.M. Subirachs, who is continuing Gaudi's main work. See the *Temple of the Sagrada Familia, Guide to the Passion of Christ Façade Sculptures by Subirachs.*

and (2) magical treatises, depicting the first seven squares related to the seven planets used in talismans. As far as type number 1 is concerned, the first known mathematical treatises on the subject were written in Arabic, from the tenth century onwards. Neither their contents nor their titles have anything to do with magic; they are purely mathematical treatises, as indicated in the title وفق الأعداد (*wafq al-aʿdād*), "Harmonic Disposition of the Numbers." In fact, only Arabic, Persian, and other Oriental treatises are known to be purely mathematical, if we reckon among them the single, extant Byzantine text, written in Greek by the scholar Moschopoulos (ca. 1330).[6] The Latin treatises, on the other hand, contain a brief mathematical description and a reproduction of the seven magic squares related to the seven planets. Examples include: *De viribus quantitates* by L. Pacioli (1498),[7] and *Practica arithmetice, & mensurandi singularis...* by G. Cardano (1539).[8] One century later, the Jesuit A. Kircher (1665) wrote the *Arithmologia,*[9] a mathematical description of the *sigilla* related to the planets, showing the same type of Mars's square introduced by Alfonso and Agrippa; while in the *Oedipus Aegyptiacus,*[10] he dealt with the seven planetary *sigilla* and the corresponding texts of talismanic magical treatises.

As for type number 2, since time immemorial and in different cultures, the combination of numbers and letters forming squares with cells under particular conditions was considered to have magical properties. However, this was not the

---

6 *Le Traité de Manuel Moschopoulos sur les carrés magiques* (ed. and transl. P. Tannery). According to Jacques Sesiano, Moschopoulos's methods were not original, but must have derived from the analysis of the squares appearing in some Arabic or Persian treatises – texts that he did not understand (Les carrés magiques de Manuel Moschopoulos, pp. 377–97, esp. 392–3); however, the expression "per three," precisely in the second building method of even squares, cannot have originated in any Arabic or Persian source, as such would have referred to the "knight's move."

7 In manuscript Bologna, Biblioteca Universitaria cod. 250, part I, chap. 90.

8 Printed in Milan, 1539; see *Caput* 42: *De propietatibus numerorum mirificis*; and 43: *De miticis numerorum propietatibus.* The magic square of order 5 in Cardano's treatise corresponds to Venus instead of Mars, because the system the author uses to order the planets is the opposite of the one habitually used. Instead of using Ptolemy's order for the fixed stars, Cardano follows Aristotle's order. According to Bouché-Leclercq in his *Histoire des Lagides*, supplemented in some areas by his later *Histoire des Séleucides*, the order of the heavenly bodies followed by Ptolemy is Saturn, Jupiter, Mars, Sun, Venus, Mercury, Moon. For the reverse order in Girolamo Cardano, cf. Láng (*Unlocked Books*, p. 92): "In Cardano, the smallest Square belongs to the Moon; while in Agrippa... the squares start with Saturn;" and Ahrens, Studien, pp. 186–250, esp. 199.

9 Printed in Rome, 1665; for the mathematical construction, see *Pronicorum Propietatibus, Caput III, De Mathematico fabrica & constructionis dictorum Sigilloum ratiocinio*, pp. 67–8.

10 Printed in Rome, 1654; see III, *Caput* VI, *Tom.* I, pp. 70–82.

case of the same numbers or letters considered individually or combined in a different form. Since ancient times, and especially amongst the Sabians of Ḥarrān, the spirits of the planets had been invoked to ask for favours,[11] mainly with magic talismans constructed during certain rituals and under specific astral conditions. For their part, the Arabs seem to have inherited the knowledge of the properties of the planets and their influences from Hellenistic astrological doctrines.[12]

The relationship of the first seven magic squares (which are those of order 3 to 9) with the magical virtues of the seven planets of Antiquity has been widely documented, especially with reference to talismanic magic.[13] In his *al-Muqaddimah,*[14] Ibn Khaldūn (fourteenth century) states that the magician (*al-sāḥir*) does not need external help, while the author of talismans (*ṣāḥib al-ṭilasmāt)* requires the aid of the spirits of the planets, the secrets of the numbers, the characteristics of the creatures, and the position of the celestial sphere which, according to the astrologers (*al-munajjimūn*), has an influence on the four terrestrial elements.[15] The name "magic squares" seems to derive from the fact that, at the beginning, the mathematical treatises were not transmitted to Latin Europe; only the magical treatises dealing with talismans related to the seven planets found their way there. The sequence of Arabic talismanic magical treatises, before reaching Europe, is as follows.

The *Rasā'il* of the Ikhwān al-Ṣafā', "The Epistles of the Brethren of Purity"[16] (tenth century) show the first seven mathematical squares in a magical and planetary environment, although not directly related to the planets. The earliest occurrence of the first seven mathematical squares related to the virtues of the seven known planets appears in Azarquiel's كتاب تدبيرات الكواكب (*Kitāb tadbīrāt al-kawākib*), or *Book on the Influences of the Planets* (eleventh century). The next treatises are those by Aḥmad b. ʿAlī al-Būnī (between the twelfth and thir-

11 According to David Pingree (Al-Ṭabari on the Prayers to the Planets, pp. 105–17, esp. 105), these rituals "are preserved in the seven *faṣls* of the third *maqāla* of the *Ghāyat al-ḥakīm*."
12 See Saliba, The Role of the Astrologer, pp. 45–68, esp. 47, n. 19.
13 The purpose of talismanic magic is to modify destiny, as opposed to "divinatory" magic, which aims to foresee the future.
14 In *A Study of History*, the British historian Arnold J. Toynbee (d. 1975) states that "the *Muqaddimah* is a philosophy of history and the greatest work of its kind that has ever yet been created by any mind in any time or place" (vol. 3, p. 322).
15 *Muqaddimah*, transl. F. Rosenthal, pp. 156 – 227. For a discussion of the reception of talismans in Greek, Muslim, and Medieval Latin milieux, see Burnett, Talismans, pp. 1–15.
16 See Marquet, Sabéens et Iḫwān al-Ṣafā'.

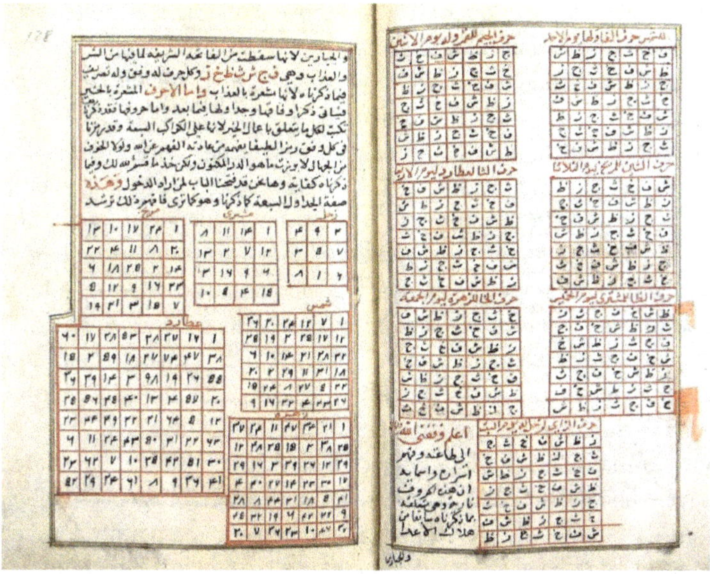

fig. 2 Aḥmad b. ʿAlī al-Būnī's *Shams al-Maʿārif* (622/1225), f. 128r

teenth centuries),[17] in which the magical use of the squares is combined with a mathematical description of their construction.

It is important to note here that the first magic square with an accompanying text related to the properties of the planets[18] found in Latin Europe is extant in the thirteenth-century Alfonsine treatise labeled by A. García Avilés as the *Libro de Astromagia.*[19] Unfortunately, the *figura Martis* with related text included in the chapter devoted to the *Formas de Mars* (ff. 27–30), is the only magic square extant in the Alfonso X treatise. It is also worthwhile recalling that under Alfonso X, also known as "the Learned," translators such as Johannes Hispalensis trans-

17 On magical squares and magic in al-Būnī's *Kitāb shams al-maʿārif* (كتاب شمس المعارف), see Ahrens, Die magischen Quadrate al-Bunis; Hermelink, Die ältesten magischen Quadrate, pp. 199–217; Lory, Magie et religion, pp. 4–15; *idem*, La magie des lettres, pp. 97–111; and Pielow, *Die Quellen der Weisheit*, pp. 4–15, esp. 7.

18 Currently being studied under the AMER Research Project mentioned above are parallels among the texts accompanying the magic squares related to the seven planets, concerning the planets' magic virtues and the influence of each one of the seven planets (Saturn, Jupiter, Mars, the Sun, Venus, Mercury, and the Moon), combined with specific elements of the animal, vegetable, and mineral world, as well as calendar and astrological planetary data which condition the favorable or unfavorable influence in the *Picatrix*, the *Clavicula Salomonis*, etc.

19 See footnote 2 above.

lated Azarquiel's treatise on magic squares: the above-mentioned *Kitāb tadbīrāt al-kawākib*, or *Book on the Influences of the Planets.*[20] To my knowledge, there are four extant manuscripts of this text, as follows (ordered by date): (1) Cairo, Dār al-Kutub, TJ 424, 3, ff. 51v–60v (twelfth–thirteenth centuries);[21] (2) Vienna, Österreichische Nationalbibliothek, 1421, ff. 2v–12v, (mid-sixteenth century); (3) London, British Library, 977, Add. 9.599, ff. 133r–145v, (seventeenth–nineteenth centuries);[22] and (4) Cairo, Dār al-Kutub, Sh124, 10 folios (access denied). The *Kitāb tadbīrāt al-kawākib* has not been studied to date, except for a three-page note in Millàs's *Estudios sobre Azarquiel*;[23] a preliminary study by Mercè Comes and Rosa Comes, *Los cuadrados mâgicos matemâticos en al-Andalus, El tratado de Azarquiel,* published in *al-Qantara* (2009);[24] and a paper, "The Arabic Tradition of Magic Squares and its transmission to Latin Europe," delivered by Mercè Comes and Rosa Comes at the congress "Contextualizing Magic."[25]

At that time, Arabic, Greek, old Castilian, and Latin treatises on talismanic magic squares were being studied in depth as part of the AMER research project. The main aim of the study is to perform a comprehensive comparative survey (codicological, paleographic, mathematic, textual, etc.) of the four groups of manuscripts dealing with talismanic magic squares and described below. The manuscripts are displayed in roughly chronological order, regardless of the language and origin.

---

20 Burnett, The Conte de Sarzana Magical Manuscript, pp. 1–7.

21 So far, we had been denied access to the two Dār al-Kutub manuscripts, perhaps due to their magical content. I am very grateful to Charles Burnett and D. Grupe for providing me with a digitalized version of MS. TJ 423 (MS. Sh124 continues to be "forbidden"). D.A. King states (*A Survey of the Scientific Manuscripts in the Egyptian National Library,* B87, 50) that there are differences between these two Cairo manuscripts, although they may likely be divergences in terms of organization; this is the case in the two other extant manuscripts, which are similar in content but arranged differently.

22 I would like to thank C. Baker, who was kind enough to indicate for us the dates of the manuscript by email on 1 April 2008.

23 Millàs, *Estudios sobre Azarquiel,* pp. 480–3.

24 Comes and Comes, Los cuadrados mágicos matemáticos, pp. 137–169.

25 Held in Rome, 3 November 2009. We there presented the first result of our research: Mercè Comes's identification of Azarquiel's *Kitāb tadbīrāt al-kawākib* as the source of the fragment of the *Libro de Astromagia*, mentioned above; see M. Comes and R. Comes, The Arabic Tradition of Magic Squares.

First Group

The first group covers the manuscripts that show exactly the same magic square for Mars as Alfonso X's treatise. The method of construction of this square in Azarquiel's *Kitāb tadbīrāt al-kawākib* (in Dār al-Kutub TJ 424.3, and British Library 977, add. 9599) and in Alfonso X's *Libro de Astromagia* is a variant derived from one of the oldest methods. This was labeled by Sesiano as "construction par placement diagonal" (by diagonals),[26] and described for the first time by Ibn al-Haytham (Iraq, 965–1041), according to an anonymous source from the eleventh century[27] – although we only know the contents through an anonymous twelfth-century manuscript.[28] This variant was labeled by Sesiano[29] as "carré oblique" and described as the "first general method in the construction of odd order squares."[30] It is also documented in the work of Moschopoulos (ca. 1300),[31] of the physician and philosopher Cornelius Agrippa (1486–1535),[32] in some "Planetenamulette,"[33] and in the work of al-Shabrāmallisī (ca. 1600),[34] *apud* al-Fullānī al-Kishnāwī (fig. 3; eighteenth century),[35] among others.

---

26 Sesiano, *Les carrés magiques dans les pays islamiques*, pp. 23–29, esp. 26, n. 34 and 256, n. 415; and Sesiano, Herstellungsverfahren, pp. 187–96, esp.190, nn. 1–2.

27 Sesiano, *Un traité médiéval*, pp. 13–15.

28 Ibid., pp. 32–35, esp. 33, n. 17.

29 Sesiano, Une compilation arabe, pp. 137–189, esp. 143–4, figs. 6–9; *idem*, *Un traité médiéval*, pp. 27–9.

30 Sesiano, Magic Squares for Daily Life, pp. 715–734, esp. 727–8, figs. 8–10.

31 Ibid., p. 382.

32 See Nowotny, The construction of certain Seals and Characters, pp. 46–57, esp. 50–51; Agrippa, *De Occulta Philosophia Libri Tres* (ed. Nowotny, 1967; and the undated Berignos Fratres print [Lyon], available at the Warburg Institute in London).

33 Wilhelm Ahrens *apud* Folkerts, Zur Frühgeschichte der magischen Quadrate, pp. 313–38, esp. 316, n. 5B.

34 Sesiano, Quelques méthodes, pp. 51–76.

35 Based particularly on al-Shabrāmallisī, ca. 1600. See Sesiano, Quelques méthodes, p. 54.

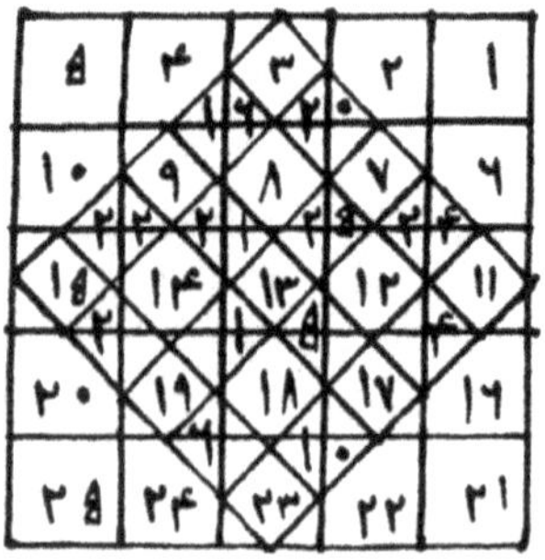

fig. 3 Square found in a copy of Shabrāmallisī's work, reproduced as is by al-Fullānī al-Kishnāwī *apud* Sesiano, *Les carrés magiques dans les pays islamiques*, p. 29, n. 38

1.1 Arabic. Azarquiel. Cairo, Dār al-Kutub TJ 424.3,[36] ff. 51v–60v, (54v); twelfth to thirteenth centuries[37]

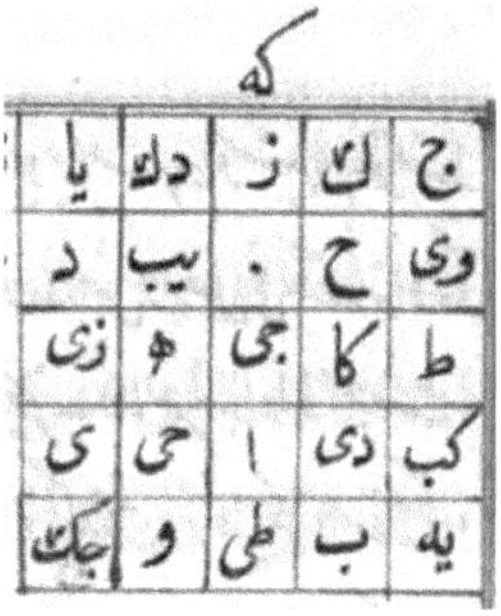

fig. 4a

25

11	*24*	7	20	3
4	12	·	8	*16*
17	5	*13*	21	9
10	*18*	1	*14*	22
23	6	*19*	2	15

fig. 4b

The square shows no errors, although some ciphers are written from right to left, and the rest (italic type in fig. 4b) from left to right. As regards the practice of leaving an empty cell and indicating the number outside the square, in this case one point marks the empty cell. As can be seen in fig. 4a, the magic square for Mars (order 5) is written in Arabic *abjad* numerals. The same happens in the

36 *Risālat fī ḥarakāt al-kawākib al-sayyāra wa tadbīrihā*, Cairo BN TJ (*Ṭalʿat majāmīʿ*) 424,3 (ff. 51v–60v, ca. 1200 H). See King, *Survey*, 50, B87, 5.2.23; and Avilés, Two Astromagical Manuscripts, p. 15.

37 Figure numbers ending with "a" correspond to the ones in the manuscript, while figures ending with "b" are transcriptions to facilitate the identification of the ciphers and errors, if any. The numbers in the square reflect the ones in the manuscript, while the corrected ones, if any, are smaller and inside brackets.

squares of order 3 and 6 in this very same manuscript, while the rest of its squares are written in the Persian variant of Eastern Hindu-Arabic notation.[38]

1.2 Old Castilian. Alfonsine scriptorium, Biblioteca Vaticana, Rome; Reg.lat.1283ª, ff. 1r–36v, (f. 27v); end of the thirteenth century[39]

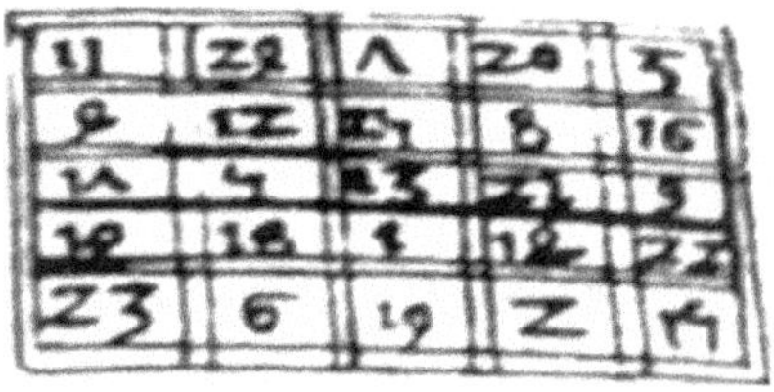

fig. 5a

11	24	7	20	3
4	12	25	8	16
17	5	13	21	9
10	18	1	14	22
23	6	19	2	15

fig. 5b

The square in this manuscript presents no errors. The manuscript, edited by A. d'Agostino[40] and later by Raquel Díez,[41] opens with an acephalous fragment in old Castilian and Western Hindu-Arabic figures.[42] The materials in the *Astromagia* came from various sources.[43] According to A. García Avilés, before reaching the Biblioteca Vaticana at the end of the seventeenth century, the manuscript could have found its way to France through the Kingdom of Catalonia and Aragon, reached England early in the fourteenth century, and, after a series of vicissitudes, returned to France, where the miscellany seems to have been put together.[44]

---

38 For the development of Eastern, Persian variant and Western Hindu-Arabic numerals, see al-Khwārizmī, *Le calcul indien* (ed. André Allard); Kunitzsch, Transmission, pp. 3–21; Burnett, Indian numerals, pp. 237–288.
39 See Avilés, Two Astromagical Manuscripts, p. 15.
40 *Astromagia* (Ms. *Reg. Lat.* 1283ª), ed. Alfonso D'Agostino, Naples: Liguori, 1992.
41 Alfonso X el Sabio: Picatrix (Ms. Vaticano Reginensis Latinus 1283a), estudio y edición, PhD diss., New York University, 1995.
42 For Western (Latin Europe) Hindu-Arabic numerals, see the sources in n. 38, above.
43 As is well known, the *Libro de Astromagia* consists of a series of folia (1r–36v) forming part of the above-mentioned miscellaneous codex kept at the Biblioteca Apostolica Vaticana. It contains twenty-one fragments of manuscripts, from the tenth to the sixteenth centuries. The subjects are astrological, juridical, historical, and literary, and the languages used are old Castilian, French, Latin, and Greek, not to mention a few marginal notes in Arabic.
44 Two Astromagical Manuscripts, pp. 20–3.

1.3 Greek. Manuel Moschopoulos, *apud* Tannery[45] and Sesiano:[46] Method "par deux et par trois;" thirteenth to fourteenth centuries

11	24	7	20	3
4	12	25	8	16
17	5	13	21	9
10	18	1	14	22
23	6	19	2	15

fig. 6

1.4 Latin. Spanish manuscript on Alchemy; Frankfurt Universitätsbibliothek, Lat. Oct. 231, ff. 133r & 140r–v, (133r); end of fifteenth century

fig. 7a

11	24	7	20	3
4	12	25	8	16
17	5	13	21	9
10	18	1	14	22
23	6	19	2	15

fig. 7b

"Zahlenexperimente" (magic square essays) in a miscellany manuscript on alchemy,[47] written in Spanish, Latin, Hebrew, and French. The author successfully constructed the magical squares of orders 3 and 5, while some others remain unfinished.

45 Manuel Moschopoulos, *Le Traité de Manuel Moschopoulos* (ed. and transl. P. Tannery), pp. 38–55.
46 Sesiano, Les carrés magiques de Manuel Moschopoulos, pp. 381–83.
47 See the *Jordanus* database (http://jordanus.badw.de/data.htm), shelfmark Lat. Oct. 231, f. 140r–v, DFRAUL8231/10.

1.5 Latin. Cornelius Agrippa;[48] *De occulta philosophia libri III*;[49] early sixteenth century

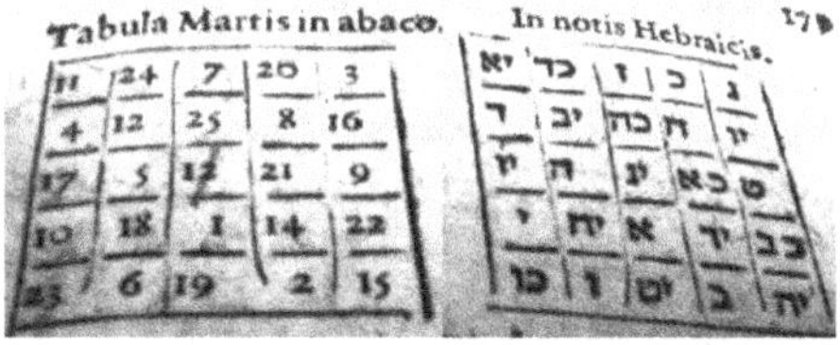

fig. 8a(1) & fig. 9a(1)

11	24	7	20	3
4	12	25	8	16
17	5	13	21	9
10	18	1	14	22
23	6	19	2	15

fig. 8b & 9b

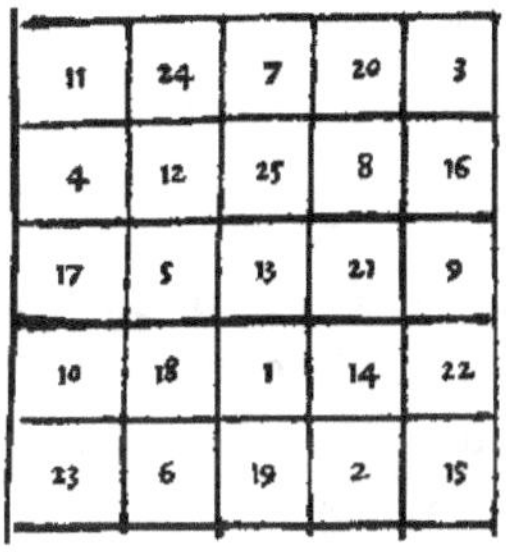

11	24	7	20	3
4	12	25	8	16
17	5	13	21	9
10	18	1	14	22
23	6	19	2	15

fig. 8a(2)

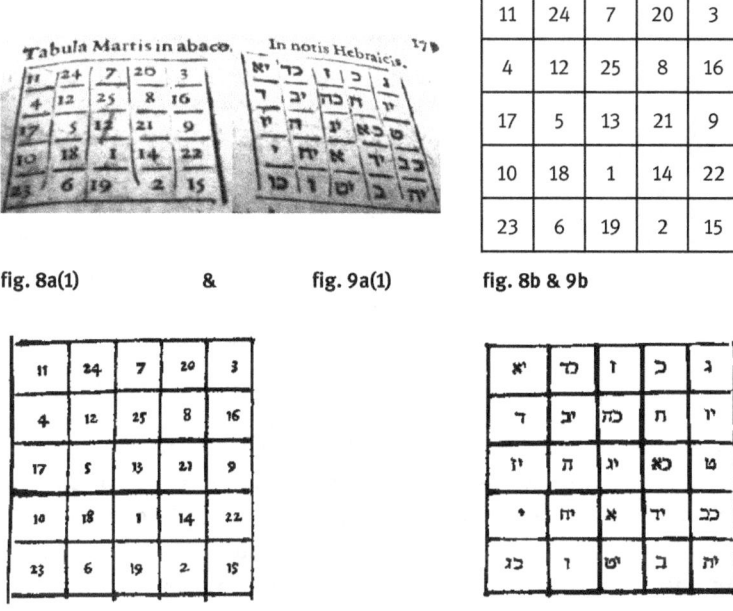

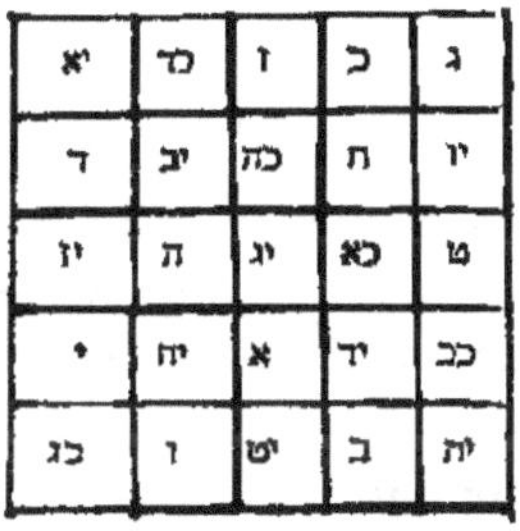

יא	כד	ז	כ	ג
ד	יב	כה	ח	יו
יז	ה	יג	כא	ט
י	יח	א	יד	כב
כג	ו	יט	ב	טו

fig. 9a(2)

Incipiunt figure septem planetarum et scias quod in istis ... De modo fabricandi figuras seu sigilla unicuique planetae proprias et conformes...

In the undated edition published in Lyon "per Berignos Fratres" (p. 179), there is a printing error in the central cell of "tabula Martis in abaco" (fig. 8a(1), above), which should read 13 instead of 12, while the correct number (13) appears "in notis hebraicis" (fig. 9a(1)). In the English edition of 1651, printed in London by Gregory Moule (reprinted by Nowotny, pp. 665–6), the squares are correct (fig. 8a(2), 9a(2)).

48 Cornelius Agrippa of Nettesheym, physician and philosopher (1486–1535). The first version of *De occulta philosophia libri III* saw the light of day in 1510, but the definitive publication including magic squares did not appear until 1533.

49 Reprinted with addenda by Karl Anton Nowotny (Agrippa, *De Occulta Philosophia Libri Tres*, pp. 665–6).

1.6 Latin. Girolamo Cardano. *Practica arithmetice & mensurandi singularis: bin quaque preter alias continentur, versa pagina demonstrabit,*[50] Chapter 42. 39; (1539).[51]

Venus.

11	24	7	20	3
4	12	25	8	16
17	5	13	21	9
10	18	1	14	22
23	6	19	2	15

fig. 10(1)

39 Et hoc dicunt multi quoniam dantur numeri planetarii & nosponemus eos verum in aſſignatione planetarum eſt conuerſus modus tenendus conuenit.n.vt plures numeri ſuperioribus tribuantur & ſunt hi.

Habēt autē comune vt ex omni latere & trāſuerſaliter eundē pſitiāt numerū.Luna 15.mercurius 34.venus 65.ſol 111.mars 175.iupiter 260.ſaturnus 350.cōueniūt etiā q nullus numerus repetitur,& q vnitatis additione pgreſſio firmatur ad quadratum,pcedūt etiā diametraliter p æqualia augumēta oēs.Quidā etiā habent in vtra q diametro vt ſaturnus ſeries cōſtitutas,vnū ē artifitio maximo talia inuēta eſſe,quorū vſus ad magiā ptinet.

fig. 10(2)

As can be seen in the text (fig. 10(2)), there is only a brief mathematical summary with a reproduction of the seven squares named for the seven known planets, but this has nothing to do with magic. The square of order 5 corresponds to Venus instead of Mars, because the system used by the author to order the planets is the opposite of the one habitually used, as we noted above.

50 Folkerts, Zur Frühgeschichte der magischen Quadrate, pp. 313 – 38, esp. 324 – 5.
51 The edition can be consulted online at http://bib.cervantesvirtual.com, and http://babel.hathitrust.org.

1.7 Latin. A. Kircher, *Oedipus Aegyptiacus*[52] (fig. 11) and *Arithmologia, sive de abditis numerorum mysteriis*[53] (fig. 12); middle of the seventeenth century.

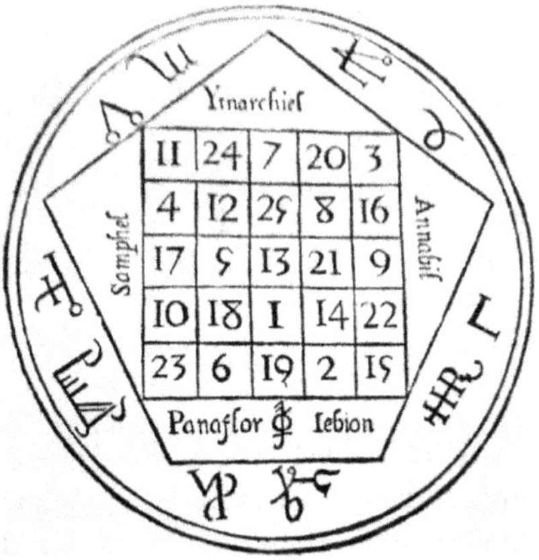

fig. 11

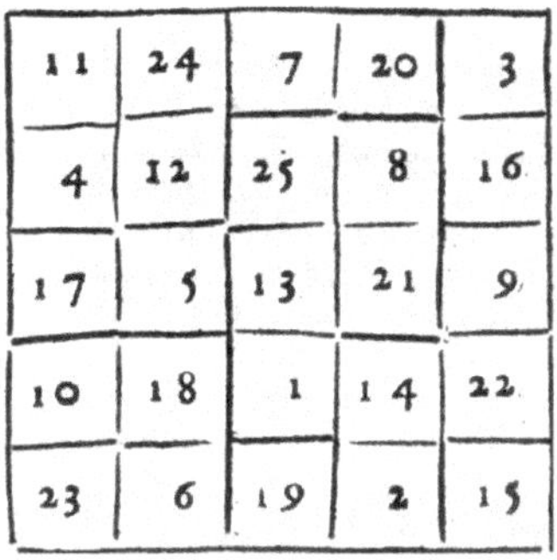

11	24	7	20	3
4	12	25	8	16
17	5	13	21	9
10	18	1	14	22
23	6	19	2	15

fig. 12

(fig. 11) *Oedipus Aegyptiacus: Dictis itaque septem planetarum arithmagicis Sigillis, septem praeparant metallorum laminas dictis planetis appropriatas... sed videamus phanatica deliramenta ... Abenmorgun Arabs alijs verbis vsum eius describit intractatus de Sigillis septem planetarum, vbi haec habet in Arabico idiomate in Latinum translata...*

(fig. 12) *Arithmologia, sive de abditis numerorum mysteriis.*

No errors appear in either version.

52 Athanasius Kircher, *Oedipus Aegyptiacus*, III, *Caput* VI, *Tom.* I, 70–82.

53 Athansius Kircher, *Pronicorum Propietatibus: Caput III. De Mathematico fabrica & constructionis dictorum Sigillorum ratiocinio. Arithmologia, sive de abditis numerorum mysteriis.*

1.8 Arabic. Azarquiel, London, British Library 977 Add. 9599, ff. 133r-145v (134v); seventeenth to nineteenth centuries

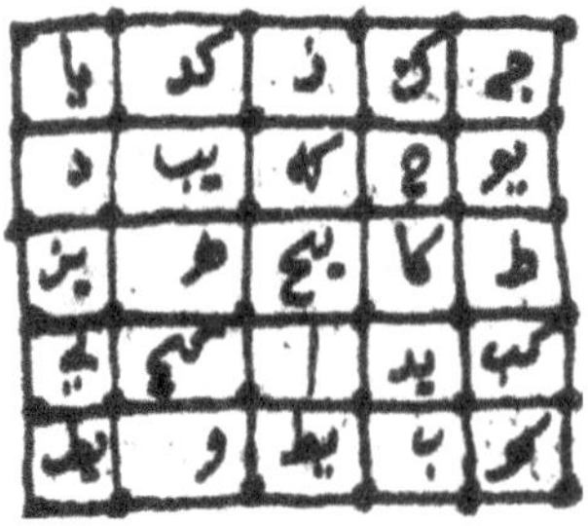

fig. 13a

11	24	7	20	3
4	12	25	8	16
17	5	(13) **18**	21	9
10	18	1	14	22
(23) **19**	6	19	2	15

fig. 13b

As can be seen in fig. 13a, the magic square for Mars in this manuscript is written in *abjad* numerals, as it is in the corresponding square in Cairo, Dār al-Kutub TJ 424.3. The ciphers are written in the expected direction – from right to left – and the squares of orders 3 and 4 present a double numeral notation in every cell, both in Eastern Hindu-Arabic and *abjad*.[54] The transcription (fig. 13b) shows that the square presents two errors: 18 instead of 13, as the result of misreading the number 3 as 8 (in the central cell), one of the most common errors in the transliteration of *abjad* alphanumerical notation (ج = 3; ح = 8)[55] together with another frequent error: the figure (19) appears twice by mistake (once in the correct place, and again in the left corner cell, which should read 23).

Second Group

The second group covers the magic squares derived from Azarquiel's *Kitāb tadbīrāt al-kawākib* preserved in Vienna.

54 For ciphers doubling the numeration in letters, in this case Eastern Hindu-Arabic numerals and Greek alphanumerical notation, see Tannery, Les chiffres arabes, pp. 355 – 60. And for the case of Greek alphanumerical notation doubled by Roman numeration, see R. Comes, Arabic, Rūmī, Coptic [etc.], pp. 157 – 185.

55 The same errors are found in the magic square of 2.1, below (Ms. Vienna, Österreichische Nationalbibliothek, Ms 162d (= Arabic Flügel 1421).

2.1 Arabic. Azarquiel, Vienna, Österreichische Nationalbibliothek, Ms. 162d (AF 1421), 1r-11v, esp. f. 5r. (963/1556); mid-sixteenth century[56]

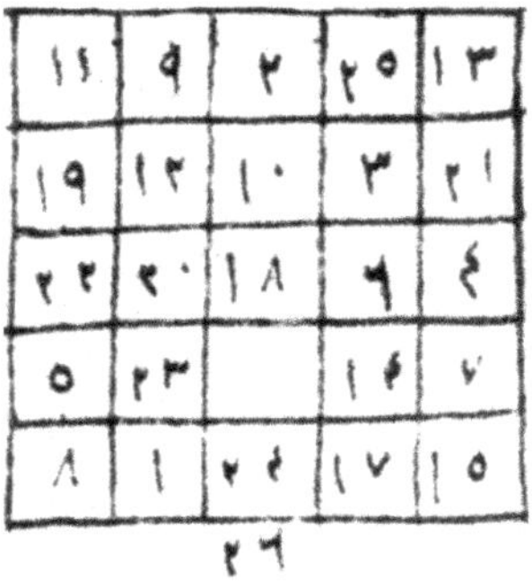

fig. 14a

11	9	2	25	(18) 13
19	12	10	3	21
22	20	(13) 18	6	4
5	23		14	7
8	1	24	17	15

(16)
26

fig. 14b

As can be seen in fig. 14a, the magic square for Mars in this manuscript is written in Eastern Hindu-Arabic numerals and starts in a quite unusual position: the second to the left of the last row. It is constructed dextrorsely, by means of the complementary diagonal method, described as a particular case of simple odd magic square method in an anonymous manuscript of the eleventh century.[57] As shown in the transcription (fig. 14b), the square presents three errors: ciphers representing 3 and 8 (corresponding to 13 and 18) are exchanged, as also occurs in the squares of order 3 and 6 of the same manuscript, which shows that at least these squares were copied from a manuscript written in *abjad*.[58] As mentioned above, this is a common error; the difference between 3 and 8 in *abjad* consists only of a point (ج = 3; ح = 8).[59]

Apart from the square of order 4, the fact that squares of this manuscript are written in Eastern Hindu-Arabic numerals suggests a contamination, or, perhaps, the hesitation of the copyist on seeing the numerical notation of the original. The remaining error, probably paleographical, coincides with the number copied outside the square; and, as usual in some Arabic manuscripts, the corresponding cell – in this case just below the central one – was left empty. This is unlike Azar-

56 Facsimile in Nowotny's edition of Agrippa, *De Occulta Philosophia Libri Tres*, Appendix VIII, pp. 657–663, esp. 659.
57 Sesiano, *Les carrés magiques dans les pays islamiques*, pp. 29–30, s.v. "4. un méthode apparentée."
58 It is also possible that the copyist, seeing that he had already written the number 13, decided to change it to 18 when reaching the central cell.
59 See Juste, Les Alchandreana primitifs (*Annex II, Alphabets numeriques*), pp. 675–82.

quiel in manuscripts Vienna 1421 and Cairo National Library, TJ 424, 3, in which case no cells are left empty. The number 26 instead of 16 is probably due to a repetition of the 2 from the number 24 in the cell just above.

Although, to my knowledge, in this tradition no other square exactly matches this one, the diagonal method used to construct it is a variant of the method developed by Ibn al-Haytham (965–1041), according to two anonymous eleventh- and twelfth-century sources,[60] and used to build the magic squares of our second group.

Third Group

The third group consists of manuscripts containing a magic square for Mars that is not exactly the same as those in groups 1 and 2, but built with a very similar system of construction;[61] this has been labelled by Camman as "diamond method,"[62] and developed by Sesiano as "Second method, by means of a rhombus"[63] or "carré oblique."[64]

60 Sesiano, *Les carrés magiques,* 23–29, esp. 26, f. 34 and 256, f. 415; Sesiano, Herstellungsverfahren magischer Quadrate aus islamischer Zeit, esp. p. 190, figs. 1–2; and Sesiano, Quelques méthodes arabes, pp. 58–59.

61 According to Jacques Sesiano, this is a method devised in the Islamic world towards the end of the tenth century or at the beginning of the eleventh, and related to the one used in the treatises of Azarquiel, Alfonso, and Agrippa (*Un traité médiéval*; *Les carrés magiques*, pp. 33–4, ff. 45–47, and 256, f. 416; and Magic Squares for Daily Life).

62 Cammann, Islamic and Indian Magic Squares, pp. 196–198.

63 Sesiano, Magic Squares for Daily Life, pp. 728–30, figs. 11–15.

64 Sesiano, Quelques constructions, pp. 251–62, esp. 253–4, figs. 6–8.

3.1 Latin. London, British Library, Harley 2404, *Liber de septem figuris septem planetarum*, ff. 60v-64v, (f.61r); fourteenth century[65]

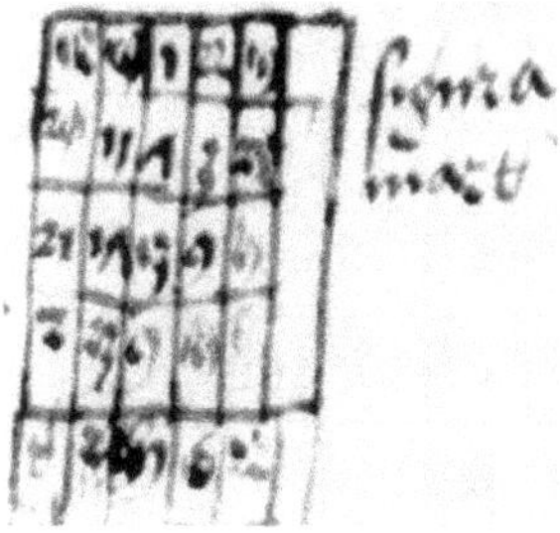

fig. 15a

14	10	1	22	(18) 19
20	11	7	3	24
21	17	13	9	5
2	23	(19) 9	15	6
8	4	25	16	12

fig. 15b

The manuscript presents two errors,[66] perhaps related to each other, since the number 19 had already been written in the first row by mistake (instead of 18), and the copyist wrote 9 where he should have written 19, without realizing that the 9 was in the upper row; or it may just have been a repetition of the 9. These are both common paleographic errors.

65 Thorndike and Kibre, *A Catalogue of Incipits*, col. 738 (14c); Klaassen, Manuscripts of Astrological Image Magic.
66 It also bears some modern miscorrections.

3.2 Latin. Oxford, Corpus Christi College Library 125,[67] Miscellany on Alchemy, ff.76r-77v, (f. 76r); late thirteenth to early fourteenth centuries[68]

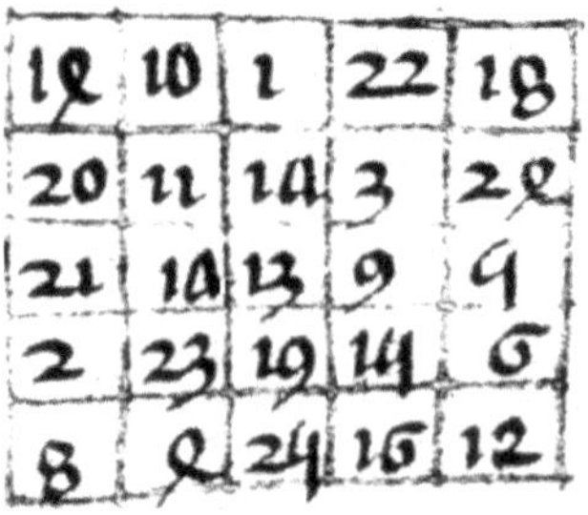

fig. 16a

14	10	1	22	18
20	11	(7) **17**	3	24
21	17	13	9	5
2	23	19	15	6
8	4	25	16	12

fig. 16b

Liber de septem figuris septem planetarum.
Incipiunt figure septem planetarum et scias quod in istis ... De modo fabricandi figuras seu sigilla unicuique planetae proprias et conformes...

There is an error in the cell just above the central one. Either the copyist added a 1 before the 7 or just anticipated the number 17, a common paleographical error (as in the Cambridge manuscript below).

67 *Tractatuum, praecipue alchemicorum, catalogus "olim peculium Thomae de Wyvelesburgh, postea Thomae Sprot, de librario S. Augustini Cantuariensis, deinde Johannis Typtofte, comitis Worcestriae, et denique Johannis Dee."*

68 Dated to the thirteenth to fourteenth centuries by Sophie Page (Magic at St Augustine's, pp. 94 – 5), according to A. Gribomont, and Menso Folkerts (Zur Frühgeschichte der magischen Quadrate, pp. 313 – 38); to the fourteenth century by Thorndike and Kibre (*A Catalogue of Incipits*, col.738; but to the early fifteenth century in Coxe, (*Catalogus Codicum*, pp. 44 – 6 [18]), and Leedham-Green and J. Roberts, *Renaissance Man.* In my opinion it is older than the Cambridge manuscript discussed in 3.3.

3.3 Latin. Cambridge University Library Add. 4087, ff. 38r-40v; fourteenth century[69]

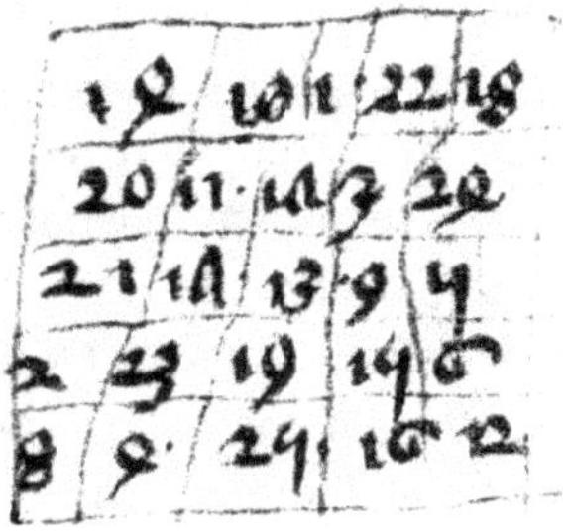

fig. 17a

14	10	1	22	18
20	11	(7) **17**	3	24
21	17	13	9	5
2	23	19	15	6
8	4	25	16	12

fig. 17b

Incipiunt figure 7 planetarum et scias quod in istis 7 figuris occultaueruntur antiqui sapientes et philosophi 7 nomina gloriosa di que sunt quatenus 7 planetas...

This magic square shows exactly the same error as in the Oxford manuscript above.

3.4 Latin. Vatican City, Biblioteca Apostolica Vaticana, Ott. 1809,[70] ff. 21r-25v, (f. 22r); fourteenth century[71]

fig. 18a

14	10	1	22	18
20	11	(7) **17**	3	24
(21) **31**	17	(13) **3**	9	5
2	23	19	15	6
8	4	25	16	12

fig. 18b

Incipiunt quaedam capitula de secretis secretorum Ptolomei scilicet de figuris et ymaginibus 7 planetarum ...

69 Thorndike and Kibre, *op. cit.*

70 Fossier (Premières recherches, pp. 381–456, esp. 429), in the "Catalogue sommaire des manuscrits latins du cardinal Marcello Cervini," records the following (no. 265): "VI, 844: Liber algorismi de integris. Ptolémée. De secretis. ? = Ottob lat 1809. (anc. Sirleto mathéma-

The square contains three errors: the first one (17 instead of 7) is the same error as the one appearing in the Oxford and Cambridge manuscripts above. Perhaps this was copied from one of them or from another of the same family. The other two errors seem to have been deliberate attempts to correct the now wrong additions (third row and third column) resulting from this error, adding and subtracting 10 units. In the central cell (3 instead of 13) the copyist probably subtracted the extra 10 units of the first mistake. This resulted in a wrong addition of the central row, which the copyist tried to correct by adding 10 units in the first cell (10 + 21 = 31). However, 31 is an impossible number for a square of order 5, since the highest number is $n^2 = 5^2 = 25$; what is more, this caused a new incorrect sum in the second row and, as a result of the second correction, the same happens in the first column. Perhaps the copyist gave up, seeing that each correction implied another one. All this shows that the copyist was unaware of the rules for constructing magic squares.

3.5 Latin. Erfurt, Wissenschaftsliche Bibliothek, CA 4º 361, f.59r-v, (59v);[72] fourteenth to fifteenth centuries[73]

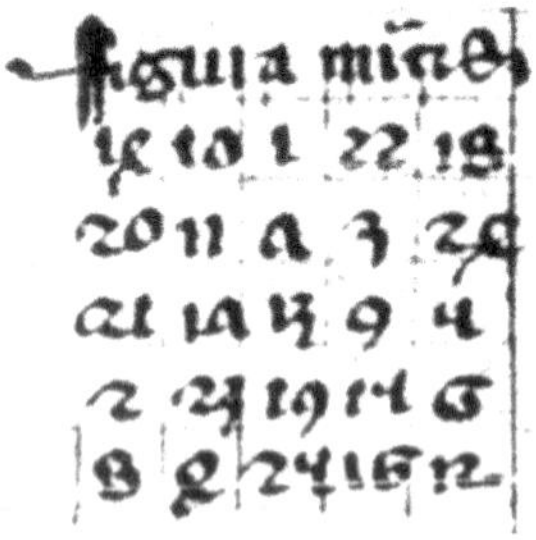

fig. 19a

14	10	1	22	18
20	11	7	3	24
21	17	13	9	5
2	23	19	15	6
8	4	25	16	12

fig. 19b

De septem figuris planetarum.
Antiqui philosophi sapientes in 7 figuris sequentibus occultauerunt 7 gloriosa nomina Dei mundi nostri accepta su(b)propietatem 7 planetarum ...

tiques 15). *V. J.* Daly et Ch. Ermatinger, *Mathematics in the codices Ottoboniani latini*, dans *Manuscripta*, t. IX (1965), p. 14."

71 *Jordanus* database, *op. cit.*

72 Probably copied in England, according to Folkerts, Zur Frühgeschichte der magischen Quadrate.

73 Fourteenth century according to Thorndike and Kibre (*A Catalogue of Incipits*, col. 110) and Folkerts (Zur Frühgeschichte der magischen Quadrate, pp. 313 – 38); but fifteenth century according to Klaassen (Manuscripts of Astrological Image Magic).

This magic square shows no errors.

3.6 Latin. Copenhagen, Kongelige Bibliotek, GKS 1658, Miscellany manuscript on Medicine, ff. 236v-240v, (228 – 240, esp. 237r);[74] fourteenth to fifteenth centuries[75]

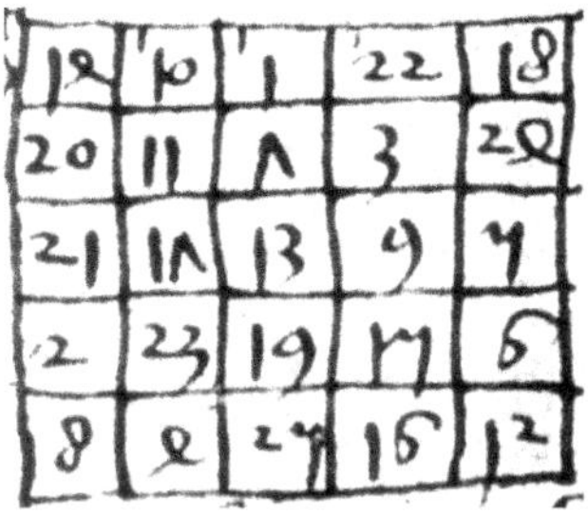

fig. 20a

14	10	1	22	18
20	11	7	3	24
21	17	13	9	5
2	23	19	15	6
8	4	25	16	12

fig. 20b

De septem figuris planetarum cum earum orationibus nec non subfumigationes Antiqui propheti et sapientes in septem figuris occultauerunt ...

This magic square shows no errors.

74 Regarding GKS 1658, see Jørgensen, *Catalogus*, pp. 437 – 9 (Gentilis de Fulgineo, *De Dosibus*).

75 Fourteenth to fifteenth century according to Thorndike and Kibre, *A Catalogue of Incipits*; but fifteenth century according to Klaassen, Manuscripts of Astrological Image Magic, and Jørgensen, *op. cit.*

3.7 Latin. Vienna, Österreichische Nationalbibliothek, Cod. 5239, ff. 147v-149r, esp. 147v;[76] fourteenth to fifteenth centuries[77]

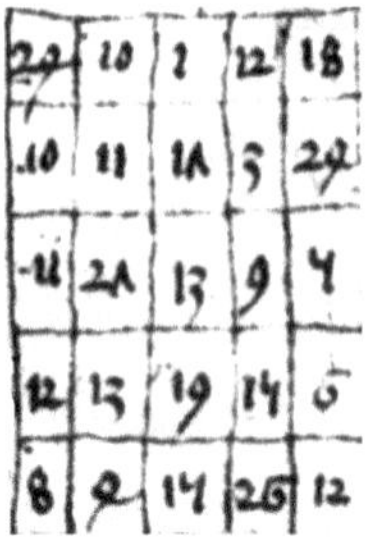

fig. 21a

(14) **24**	10	1	(22) **12**	18
(20) **10**	11	(7) **17**	3	24
(21) **11**	(17) **27**	13	9	5
(2) **12**	(23) **13**	19	15	6
8	4	(25) **15**	(16) **26**	12

fig. 21b

De modo fabricandi figuras seu sigilla unicuique planetae proprias et conformes. Incipiunt figure 7 planetarum et scias quod in istis 7 figuris occultauerunt philosophi et sapientes antiqui 7 Dei nomina gloriosa que sunt super 7 planetas ...

This manuscript presents ten errors. One of them, 17 instead of 7, indicates that either this manuscript was copied from the above manuscripts or from a common source. The remaining nine errors show the same type of correction (adding and deducting 10 units) as we find in the Vatican manuscript above – although, in this case, the copyist managed to make all the columns and rows add up to 65. This does not mean that the magic square had been correctly constructed, especially considering that the figures 26 and 27 can never appear in a square of order 5 (highest number = 25), which shows again that the copyist was unaware of the rules for constructing magic squares.

76 There is a good reproduction in Karl Nowotny's reprint of Agrippa, *De Occulta Philosophia Libri Tres* (Appendix IX, pp. 665 – 6); see also Thorndike and Kibre, *A Catalogue of Incipits*, col. 738.

77 Cf. the date in Sesiano (Magic Squares for Daily Life, pp. 715 – 34), who offers an almost complete edition and translation into English. See also Nowotny's reprint (*op. cit.* and p.906); Österreichische Nationalbibliothek, *Tabulae*, pp. 70 – 71; and Thorndike and Kibre, *A Catalogue of Incipits*, col. 738.

3.8 Latin. Cambridge University Library, Ms. Dd. XI 45, Herbal and Medical Miscellany, ff. 138r–139v (138v): fourteenth to fifteenth centuries[78]

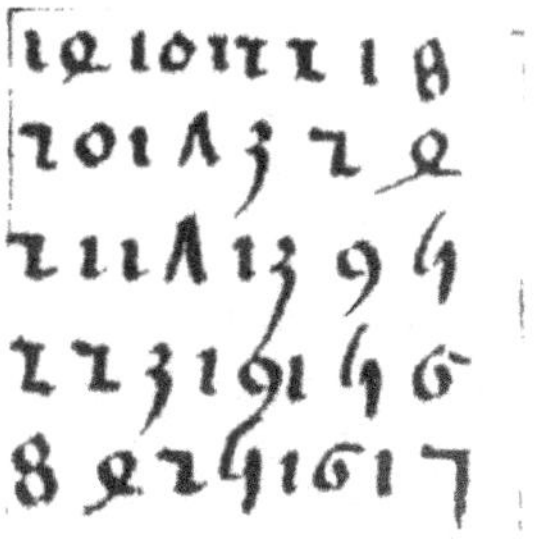

fig. 22a(1)

14	10	(1) 12	(22) 2	18
20	(11) 17	(7) 3	(3) 2	(24) 4
21	17	13	9	5
2	23	19	15	6
8	4	25	16	12

fig. 22b(1)

11	24	7	20	3
4	12	25	8	16
17	5	13	21	9
10	18	1	14	22
23	6	19	2	15

fig. 22a(2) Square of type 2
Square added by Lidaka to the edition, to replace the wrong one in the manuscript.

11	24	7	20	3
4	12	25	8	16
17	5	13	21	9
10	18	1	14	22
23	6	19	2	15

fig. 22b(2)

Liber de Angelis, Liber de ymaginibus planetarum. Secreta astronomia de sigillis planetarum & eorum figuris.

Seeing that the squares appearing in the manuscript had errors, the editor of the edition of this text inexplicably "replaced the squares with more readily understood squares" (sic!)[79] With the text accompanying Mars, he used the type of square corresponding to our second group, while copying the actual square appearing in the manuscript on p. 66, n. 47. The original square shows six errors, all of which could be paleographic – although this does not seem to be the kind of confusion that arises when copying *abjad* numerals, nor does there appear to be any relationship among them.

78 A record may be found online at Late Medieval English Magic, https://magicalmedieval.wordpress.com/2014/04/30/culddxi45.

79 Lidaka, The Book of Angels, pp. 32–75, esp. 36 and 66–7.

3.9 Latin. Krakow, Biblioteka Jagiellońska, BJ 793, ff. 60r-61r (60r); (ca. 1459)[80]

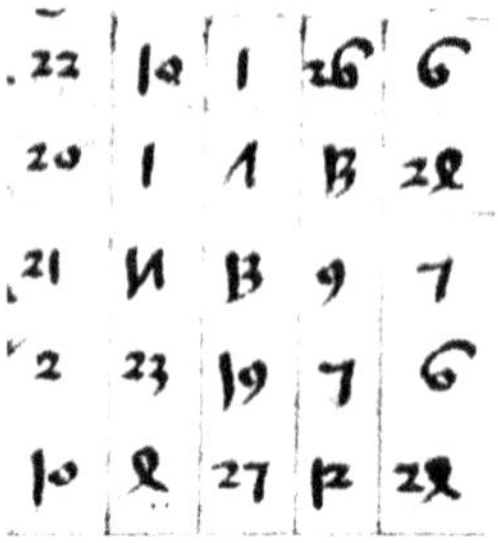

fig. 23a

65	(14) **22**	10	1	(22) **26**	(18) **6**
65	20	(11) **1**	7	(3) **13**	24
	21	17	13	9	5
	2	23	19	(15) **5**	6
75	(8) **10**	4	25	(16) **12**	(12) **24**

fig. 23b

This square presents nine errors: the first and second rows and the fourth and fifth columns show multiple attempted corrections (adding and deducting different quantities). However, the 26 in the first row is impossible (highest num. = 25) and the 10 in the first column and last row could only have been an impossible 0 in order to add up to the quantity required: 65. Once again, the copyist was unaware of the methods of construction.

80 Rosińska, Scientific Writings.

3.10 Latin. Marburg, Universitätsbibliothek, Mscr., 18, ff. 50r-53r, (50v); (1479)[81]

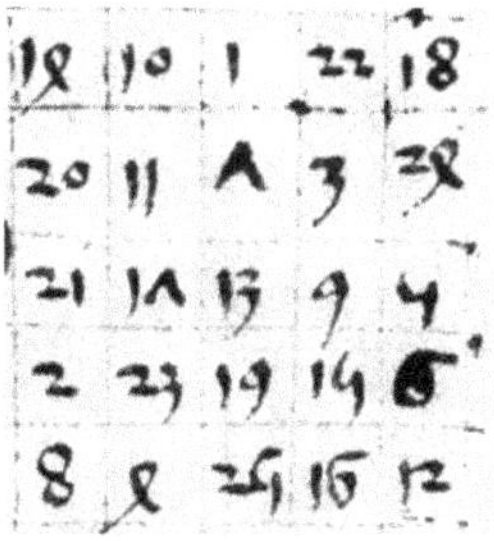

fig. 24a

14	10	1	22	18
20	11	7	3	24
21	17	13	9	5
2	23	19	15	6
8	4	25	16	12

fig. 24b

De septem figuris planetarum.
Antiqui sapientes philosophi in 7 figuris sequentibus occultaverunt... propietatis figuris gloria dei accepta planetarum qui sunt Saturnus Iupiter, mars, Sol, uenus, mercurius, luna[i] *quia sunt eorum ... nomina traduxentur figura*

This magic square shows no errors.

3.11 Luca Pacioli,[82] *De viribus quantitatis*, "codice 250 della Biblioteca Universitaria di Bologna," I, f. 18v (1498)[83]

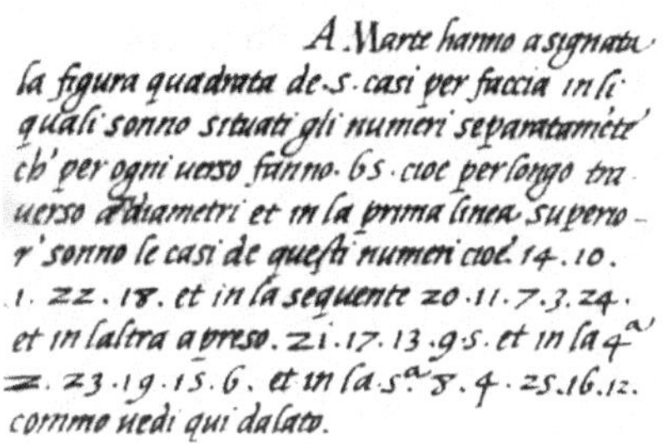
A Marte hanno asignata la figura quadrata de 5 casi per faccia in li quali sonno situati gli numeri separatamente ch' per ogni uerso fanno 65 cioe per longo traverso et diametri et in la prima linea superior sonno le casi de questi numeri cioe 14.10.1.22.18. et in la sequente 20.11.7.3.24. et in laltra apreso. 21.17.13.9.5. et in la 4a 2.23.19.15.6. et in la 5a 8.4.25.16.12. commo uedi qui dalato.

fig. 25a

14	10	1	22	18
20	11	7	3	24
21	17	13	9	5
2	23	19	15	6
8	4	25	16	12

fig. 25b

81 Heyne, *Die mittelalterlichen Handschriften*, 45 – 6.
82 Luca Pacioli, mathematician (ca. 1446 – 1517).
83 Translated by Augusto Marinoni (*Introduction to Luca Pacioli's De Viribus Quantitatis*).

The squares are not drawn, but described in words: the author indicates the numbers corresponding to each row, as can be seen in fig. 25a. The *figura quadrata* is related to the planets, without magical purposes.

3.12 Latin. Wolfenbüttel, Herzog August Bibliothek, Guelf 17–8Aug, ff. 186r-190r, (187v); early sixteenth century[84]

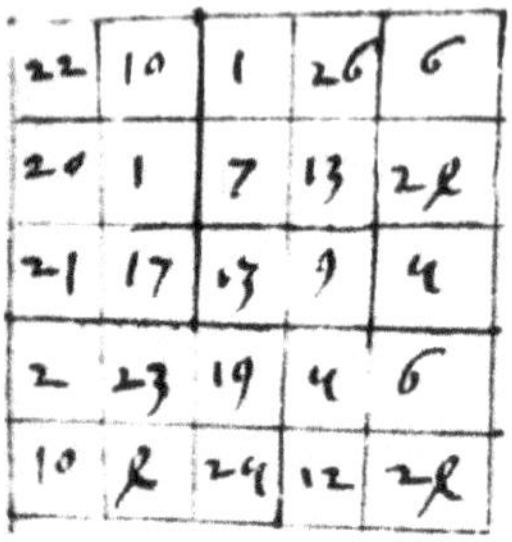

fig. 26a

(14) 22	10	1	(22) 26	(18) 6
20	(11) 1	7	(3) 13	24
21	17	13	9	5
2	23	19	(15) 5	6
(8) 10	4	25	(16) 12	(12) 24

fig. 26b

Sigilla sive figure planetarum scripta in.

Here we find exactly the same nine errors as occur in 3.9, above. Therefore, either Krakow BJ 793 is the source for this manuscript, or both share the same source.[85]

3.13 Greek. Oxford, Bodleian, Holkham Greek 109, ff. 7–9v;[86] fifteenth century[87]

14	10	1	22	18
20	11	7	3	24
21	17	13	9	5
2	23	19	15	6
8	4	**25**	**16**	**12**

fig. 27

84 von Heinemann, *Die Augusteischen Handschriften*, no. 3098, p. 209.
85 This will be investigated in a forthcoming critical edition.
86 Barbour, Summary Description.

There are no errors in this manuscript's square, according to A. Gribomont.[88]

3.14 English. British Library, Ms. Sloane 3826, *Liber Lunae*, ff. 93r-96v (94r); sixteenth century[89]

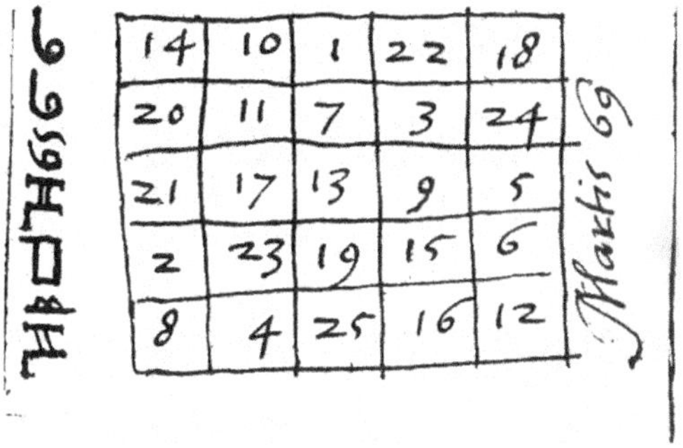
14 10 1 22 18
20 11 7 3 24
21 17 13 9 5
2 23 19 15 6
8 4 25 16 12
Martis 69

fig. 28

In the name of the meeke God and mercifull, to God alone honor and glory. This is liber... that is said Liber Lunae ...Now followeth the figure of the Planets

This square shows no errors.

---

87 To be more precise: 1470–1480, according to Colinet, et al., eds. *Les Alchimistes Grecs*, Tome XI, p. xvii.

88 A. Gribomont, email dated 10 May 2011. I have not yet been able to see the manuscript myself.

89 The *Liber Lunae* forms part of a magical compendium, referred to as *Liber Salomonis* or *Cephr Raziel*, covering ff. 84r-97v; the magic squares of the 7 planets are in ff. 93r-96v. The British Library, Sloane 3826 MS of the *Liber Lunæ:* 84r–100r has been transcribed, edited, and introduced by Don Karr (http://www.digital-brilliance.com/contributed/Karr/Solomon/).

3.15 Latin. London, Wellcome Institute for the History of Medicine 128, *De figuris planetarum*, ff. 18v-21r (19v) (1487)[90]

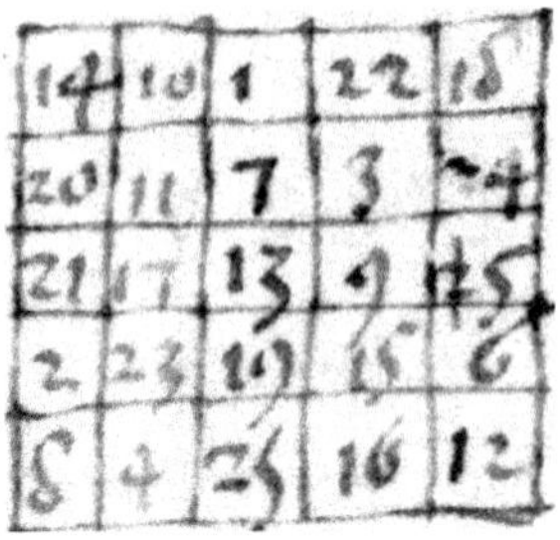

fig. 29a

14	10	1	22	18
20	11	7	3	24
21	17	13	9	5*
2	23	19	15	6
8	4	25	16	12

fig. 29b

De septem Figuris planetarum. Iste sunt figure ·7· planetarum in quibus conduntur dei nomina quae sunt super planetas quae non licet nominare.

There is one error (25 instead of 5) – but it has been corrected with the same ink and hand.

3.16 Latin. Ghent, Universiteitsbibliotheek Hs 1021 A, ff. 67v-78v, (f. 70v–73r); fifteenth to sixteenth centuries[91]

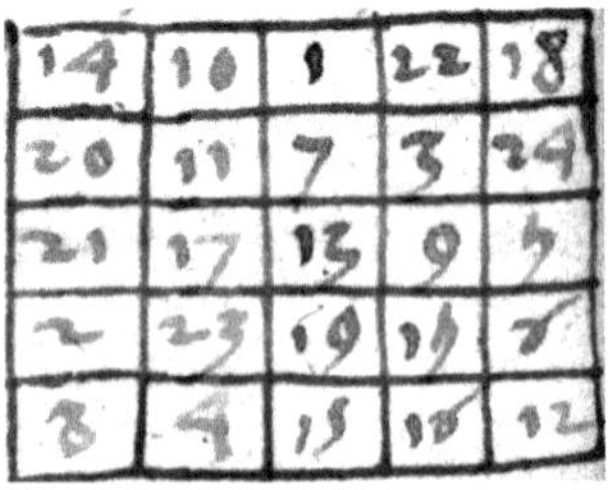

fig. 30

Geberi Magi et philosophi expertissimi De fuguris septem planetarum per numeros in quibus occultauerunt semtem [sic] nomina Dei.
Prefatio: Antiqui sapientes et philosophi in astronomia et magis experti in septem figuris seguentibus occultauerunt septem nomina gloriosi dei, qui posuerunt super proprietates septem planetarum

90 The folia preceding this treatise (ff. 14–17) cover extracts of the Picatrix. See al-Majrīṭī, *Picatrix* (Pingree, ed.), pp. LXII–LXVI.

91 Dated tentatively to the late fifteenth or early sixteenth century, due to rather modern writing and ciphers (Braekman, *Middeleeuwse*, p. 506).

This square shows no errors.

3.17 Latin and German.[92] Pseudo-Paracelsus, pp. 131–138 (134);[93] (probably 1525)[94][95]

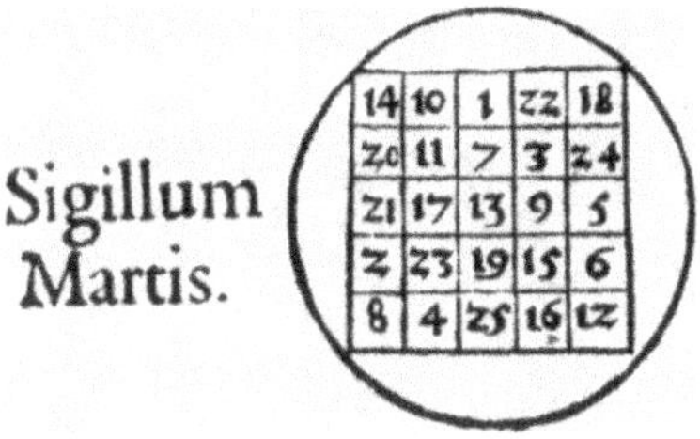

fig. 31(1)

fig. 31(2)

92 The text in Nowotny's Agrippa is in German, although it was translated into Latin by A. Schröter and published in Krakow in 1569, according to Szőnyi, Occult Sciences.

93 Pseudo-Paracelsus, *Liber Septimus, Archidoxis Magicae: De sigilis planetarum.* See the text and images reproduced in Nowotny's reprint of Agrippa's *De occulta philosophia,* Appendix XIb, pp. 677–680, esp. 679.

94 According to Schneider, *Paracelsus*, apud Pagel, [Review], p. 331.

95 This image corresponds to Hiebner, *Mysterium Sigillorum*, pp. 140–147, esp. 144. Cf. the text and images reproduced in Nowotny's reprint of Agrippa, *De occulta philosophia*, Appendix XIIa, pp. 681–683, esp. 682.

3.18 A number of medal-talismans from the seventeenth century[96]

fig. 32

All of these squares were constructed using the following interrelated methods of construction: the complementary diagonal method (pandiagonal); the diamond method;[97] or the oblique square method.[98] This latter is described as the third method for constructing odd squares in two anonymous treatises kept in Istanbul,[99] although, according to V. Karpenko, it could have been constructed by means of the Persian method for the construction of odd-order squares, in which "the 1 is placed immediately below the centre cell and the following numbers are written either diagonally or in positions corresponding to the movements of a knight in chess." [100] Fig. 33 shows some instances of fully developed methods in British Library, Persian Ms. Add 7713.[101]

96 This image corresponds to Reichelt, *Exercitatio*, Tabula III. Cf. the images reproduced in Nowotny's reprint of Agrippa, *De occulta philosophia*, Appendix XVI, pp. 715 – 718, esp. 716.
97 Cammann, Islamic and Indian Magic Squares, pp. 181 – 209, esp. 197, figs. 4c–d.
98 Sesiano, *Un traité médiéval*, pp. 32 – 35, fig. 7; *idem*, Quelques constructions, pp. 251 – 62, esp. 253 – 4, figs. 6 – 8; *idem*, Quelques méthodes, pp. 58 – 59; and *idem*, Quadratus Mirabilis, pp. 199 – 233.
99 Sesiano, *Un traité médiéval*, pp. 12 – 15, n. 15.
100 Karpenko, Two Thousand Years, pp. 147 – 53, esp. 147.
101 *Cabalistic Work*, ff. 50r–50v, in Rieu, *Catalogue*, vol. 2, p. 487.

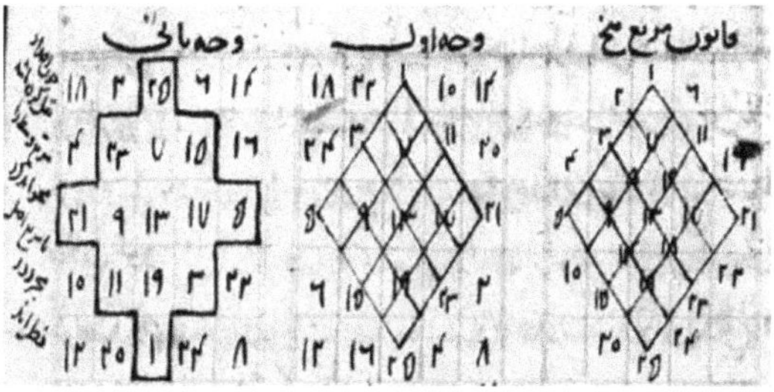

fig. 33

Fourth Group: Other Magic Squares Designated for Mars[102]

21	3	4	12	25
15	17	6	19	8
10	24	13	2	16
18	7	20	9	11
1	14	22	23	5

fig. 34 Ikhwān al-Ṣafā' *apud* Hermelink (Die ältesten magischen Quadrate, p. 208) and M. Souissi (Ḥisāb al-wafq, pp. 27–43)

102 While I was finishing this article, I heard the sad news that Professor J. Vernet had just passed away. I will always be grateful to him for his advice and support, and will never forget the day when, knowing I was interested in magic squares, he handed me the documents that he had assembled during his years of study of this subject, including this Arabic publication by M. Souissi, together with Vernet's own handwritten translation.

23	20	12	9	1
7	4	21	18	15
16	13	10	2	24
5	22	19	11	8
14	6	3	25	17

fig. 35 Moschopoulos *apud* Hermelink (Die ältesten magischen Quadrate, p. 209)

10	18	1	14	22
4	12	25	8	16
23	6	19	2	15
17	5	13	21	9
11	24	7	20	3

fig. 36 Moschopoulos apud Sesiano, method "par trois et par 5" (Les carrés magiques de Manuel Moschopoulos, pp. 377–97, esp. 383–5, fig. 7); Sesiano (Construction of Magic Squares, pp. 1–20, esp. 5–6, fig. 10); and P.G. Brown (The Magic Squares of M. Moschopoulos)

17	24	1	8	15
23	5	7	14	16
4	6	13	20	22
10	12	19	21	3
11	18	25	2	9

fig. 37 "Methode der Inder" diagonals (Ahrens, Studien, pp. 186–250, esp. 196)

fig. 38 Some more examples from British Library, Persian Ms. Add 7713, f. 60v

This study is still in progress. Among the manuscripts that I have not been able to see, I would mention the following:

- Conte de Sarzana, Private Library, Unnumbered, Geber, pp. 98–130, (1510): *De operatione figurarum septem planetarum Incipit liber ocultus et pretiosus Geberi philosophi in astronomia et magicis expertissimi de operatione figurarum admirabili septem planetarum quem transtulit Johannes Hispaniensis. Antiqui sapientes et filosophi in astrologia et magia experti in septem figuris sequentibus occultaverunt septem nomina gloriosi dei....*[103]
- Ms. Erfurt, Amploniana- Math. 54: *Liber de septem figuris septem planetarum et earum oracionibus* (sic!) *necnon subfumigacionibus* (sic!). The manuscript "heute nicht mehr existiert" according to Dr. Brigitte Pfell from Erfurt / Gotha Universitätsbibliothek, although it appears in the catalogue.[104]
- German illustrated manuscript on paper, in German, Latin, and Hebrew, from a private collection (Europe):[105] *Alchemy: Seals of the Seven Planets,*

---

103 See Burnett, The Conte de Sarzana Magical Manuscript. *Hispaniensis* is a common variant for *Hispalensis* (of Seville) which is probably the true name of the translator. A Spanish (Christian or Moorish) origin for the translation or composition of this text is also suggested by an instruction in this copy of the text (but omitted in Cambridge University Library, MS Dd. XI.45) to make a round plate from six ounces of gold in the shape of a morabotinus. This gold coin was introduced by the Almoravids into Spain, and was subsequently put into circulation by Alfonso I, Ferdinand II and Alfonso VIII of Castile. The author suggests that the sixteenth-century copy "was most probably written in Italy."

104 Schum, *Beschreibendes* (see *De Mathematica*, pp. 798–808, esp. 806, n. 54).

105 Currently for sale at http://www.textmanuscripts.com. The description reads: "Manual for the composition of talismans or seals perhaps for use by a physician-alchemist. Its pages present a combination of astrological seals with diagrammatic talismans in Greek and Hebrew letters along the lines of those created by Agrippa and Paracelsus. Unidentified alchemist, who correct-

ff. 2v-3r, Title, "De sigillo Martis;" *incipit*, "Dieses Sigill ist genommen ex sphera quinta Macheim genandt, soll gemacht werden in seinem Tag und Stunde, gegossen ex venere Martis ... mit Sandel und Pfeffer." The magic square for Mars is on f. 3r. Late sixteenth century.

- Latin and Greek. *De Plantis Duodecim Signis et Septem Planetis Subiectis; On the Signs of the Twelve Plants and on the Subjects of the Seven Planets.*[106] Primary source text: 3 pages. Origin: Spain.

Conclusion

This is a summary of the state of the study of the transmission of talismanic magic squares treatises to Latin Europe. This preliminary survey compares only the magic square of Mars, because it is the only one surviving in Alfonso X's treatise. However, to definitively establish the links among the surviving traditions, the next step (on which I am already working) is to revise and compare the rest of the squares (especially orders 4, 6, 7, 8, and 9) and their methods of construction – both inside each treatise and among the different treatises. The corresponding texts must also be compared, especially regarding: rituals; specific elements of the animal, vegetable, and mineral world; Arabic coins; calendar and astrological planetary data which condition favorable or unfavorable influence; authors or translators; *incipit* and *explicit*; etc.

Although the survey is far from being completed, it points to the fact that the Alfonsine translation of Azarquiel's treatise reached Central and Northern Europe quite late – possibly while the manuscript was in an unknown location after the dispersal of the Duke of Bedford's collection on his death in 1435, but before its appearance in the collection of Pierre Daniel d'Orleans (1530–1609) – and was not at the basis of European traditional treatises on magic squares, apart from Agrippa's work of 1533. On the other hand, it is interesting to point out that most Latin treatises showing a relationship with mathematics correspond to the first group of squares and seem to have an Italian or Spanish origin.[107] The third group, instead, seems to have been copied in England and

ed several of the figures." The annotations translate some Latin expressions, leading us to believe that the original owner who used the manuscript was not well versed in Latin.

106 Rodriguez, ed., Dos Fragmentos Inéditos, pp. 369–82.

107 Cf. MS Frankfurt, Lat. Oct. 231 "Zahlenexperimente;" Cardano's *Practica arithmetice*; and Sloane 3826 *Liber Lunae*. Alfonso's Reg. Lat. 1283 (group 1) also contains a *Liber Lunae*.

Northern Europe, even though one of the two treatises related to mathematics is of Italian origin.[108]

The information we have on the German illustrated manuscript *Alchemy: Seals of the Seven Planets,* ff. 2v-3r, is not clear regarding the group to which it pertains: "talismans in Greek and Hebrew letters along the lines of those created by Agrippa (first group) and Paracelsus (third group)" – but we know that "the unidentified alchemist corrected several of the figures." Regarding Conte de Sarzana,[109] I have not yet been able to see this last manuscript, but at the time of writing I anticipate receipt of the corresponding folia.[110] At any rate, it seems a very important link in the transmission, as it was translated in Spain by Johannes Hispalensis.

According to Sesiano, the "user did not have to be familiar with the general methods for constructing magic squares; all he needed was the knowledge of isolated squares and their properties." He continues: "That is why the accompanying texts rarely if ever give the slightest hint as to how the magic square depicted had been obtained."[111] The magical texts that reached Europe nevertheless gave rise to the efforts of some copyists to construct the squares by themselves, as witnessed by the attempts we see in some manuscripts.[112]

Another observation is that there may have been contamination from different sources, although most of them keep traces of the Arabic original, like the expression "scias quod,"[113] or the misunderstanding of some numbers, like 3 and 8, that betray an Arabic original using *abjad* alphanumerical notation.

Furthermore, it is also striking how similar our texts are to *Picatrix Latinus*, Liber III, *Capitulum septimum: "De attractione virtutis planetarum, et quomodo loqui possumus cum eis, et quoomodo effectus dividitur per planetas, figuras, sacrificia, orations, sufumigaciones, proposiciones, et status celi necessarii cuilibet*

108 L. Pacioli, *De viribus quantitatis*, presenting a brief mathematic summary. The other is the manuscript Wolfenbüttel, Herzog August Bibliothek, Guelf 17 – 8 Aug, presenting unsuccessfully constructed magical squares.

109 Burnett, The Conte de Sarzana Magical Manuscript.

110 Charles Burnett has kindly agreed to send them to me (email of 11 August 2011).

111 Sesiano, Magic Square for Daily Life.

112 For instance, in the above mentioned Lat. Oct. 231, preserved in Frankfurt, but originating in Spain in the fifteenth century; or in the Wolfenbüttel, Herzog August Bibliotek, 3098 (17 – 8. Aug.4º), of the early sixteenth century.

113 Corresponding to the Arabic أنه اعلم . See Vienna 5239, Cambridge University Library Add. 4087, Oxford, Corpus Christi College 125 (all of the 3rd group); and Vienna 1421 (of the 2nd group).

planetarum,"[114] while in Liber IV, ii, 9,[115] magical squares are mentioned and related to the proper planet, although without the *figurae*.

In short, much work is still to be done to reach reliable conclusions. Even though the magic squares of Mars point to two main traditions – as M. Folkerts already anticipated[116] – the survey of the rest of the squares (esp. of orders 4, 6, 7, 8, and 9), together with the corresponding texts, may still offer us interesting surprises.

Bibliography

Henricus Cornelius Agrippa ab Nettesheym, *De Occulta Philosophia Libri Tres*, ed. Karl Anton Nowotny, Graz: Akademische Druck- und Verlagsanstalt, 1967; reprint of Cologne: Soter, 1533.

Henricus Cornelius Agrippa ab Nettesheym, *De Occulta Philosophia Libri Tres*, London: Gregory Moule, 1651.

Henricus Cornelius Agrippa ab Nettesheym, *De Occulta Philosophia Libri Tres*, Lyon: Berignos fratres, s.d.

W. Ahrens, Die magischen Quadrate al-Bunis, *Der Islam* 12, 1922, pp. 157–77; and 14, 1925, pp. 104–10.

W. Ahrens, Studien über die magischen Quadrate der Araber, *Der Islam* 7, 1916, pp. 186 –250.

Alfonso X el Sabio, *Astromagia* (Ms. *Reg. Lat.* 1283ª), ed. Alfonso D'Agostino, Naples: Liguori, 1992.

A.G. Avilés, Two Astromagical Manuscripts of Alfonso X, *Journal of the Warburg and Courtauld Institutes* 59, 1996, pp. 14–23.

R. Barbour, Summary Description of the Greek Manuscripts from the Library at Holkham Hall, *Bodleian Library Record* 6/5, 1960, pp. 591–613.

Auguste Bouché-Leclercq, *Histoire des Lagides*, 4 vols., Paris: E. Leroux, 1903–07.

Auguste Bouché-Leclercq, *Histoire des Séleucides, 323–64 avant J.-C.*, 2 vols., Paris: E. Leroux, 1913–1914.

W.L. Braekman, *Middeleeuwse witte en zwarte magie in het Nederlands taalgebied*, Ghent: Koninklijke Academie voor Nederlandse Taal- en Letterkunde, 1997.

P.G. Brown, The Magic Squares of M. Moschopoulos, online via the Mathematical Association of America's Mathematical Sciences Digital Library (MathDL), http://www.maa.org/publications/periodicals/convergence/the-magic-squares-of-manuel-moschopoulos-introduction.

114 al-Majrīṭī, *Picatrix* (Pingree, ed.), III, vii, 4; III, vii, 11; III, vii, 23.

115 al-Majrīṭī, *Picatrix* (Pingree, ed.), IV, ii, 9: "cum **Luna** fuerit in Scorpione... et facias figuram quadratam in terra et ipsammm cooperies foliis nucum et foliis malorum citoniorum et de penolis; subriga totum cum aqua rosacea...ponas ante te **novem** (the magic square of the moon is of order 9) thuribula argentea... lignum aloes, storacem et incensum... pannis albissimis ... ."

116 Folkerts, Zur Frühgeschichte der magischen Quadrate.

C. Burnett, The Conte de Sarzana Magical Manuscript, in *Magic and Divination in the Middle Ages: Texts and Techniques in the Islamic and Christian Worlds*, Aldershot, Great Britain; Brookfield, Vt., USA : Variorum, 1996, art. IX, pp. 1–7.

C. Burnett, Indian Numerals in the Mediterranean Basin in the Twelfth Century, with Special Reference to the "Eastern Forms," in Yvonne Dold-Samploniusm, Joseph W. Dauben, Menso Folkerts, and Benno van Dalen, eds., *From China to Paris: 2000 Years Transmission of Mathematical Ideas*, Stuttgart: Franz Steiner Verlag, 2002, pp. 237–288 [reprinted in: C. Burnett, *Numerals and Arithmetic in the Middle Ages*, Variorum Collected Studies Series CS967, Aldershot: Ashgate, 2010, art. V].

C. Burnett, Talismans: Magic as Science? Necromancy among the Seven Liberal Arts, in *Magic and Divination in the Middle Ages: Texts and Techniques in the Islamic and Christian Worlds*, Aldershot, Great Britain; Brookfield, Vt., USA : Variorum, 1996, art. I, pp. 1–15.

S. Cammann, Islamic and Indian Magic Squares (I), *History of Religions* 8/3, 1969, pp. 181–209.

Girolamo Cardano, *Practica arithmetice, & mensurandi singularis* (*Caput* 42: *De propietatibus numerorum mirificis*; and 43: *De miticis numerorum propietatibus*), Milan: Bernardini Calusci, 1539.

A. Colinet, et al., eds. *Les Alchimistes Grecs*, Tome XI, Paris: Belles Lettres, 2010.

M. Comes and R. Comes, Los cuadrados mágicos matemáticos en al-Ándalus, El tratado de Azarquiel, *Al-Qantara* 30, 2009, pp. 137–169.

M. Comes and R. Comes, The Arabic Tradition of Magic Squares. The First Steps of its Transmission to Latin Europe, in Marina Piranomonte and Francisco Marco Simón, eds, *Contesti Magici*, Roma: De Luca, 2012, pp. 327–334.

R. Comes, Arabic, Rūmī, Coptic or Merely Greek Alphanumerical Notation? The Case of a Mozarabic 10th Century Andalusī Manuscript, *Suhayl (Journal for the History of the Exact and Natural Sciences in Islamic Civilisation)* 3, 2003, pp. 157–185.

H.O. Coxe, *Catalogus Codicum MSS.*, Pars II *Collegii Corporis Christi*, Oxford: Typogr. Acad., 1852.

R. Díez, Alfonso X el Sabio: Picatrix (Ms. Vaticano Reginensis Latinus 1283a), estudio y edición, PhD diss., New York University, 1995.

M. Folkerts, Zur Frühgeschichte der magischen Quadrate in Westeuropa, *Sudhoffs Archiv, Zeitschrift für Wissenschaftgeschichte*, B. 65, 1981, pp. 313–38.

F. Fossier, Premières recherches sur les manuscrits latins du cardinal Marcello Cervini 1501–1555, *Mélanges de l'École française de Rome, Moyen-Âge, Temps modernes* 91/1, 1979, pp. 381–456.

O. von Heinemann, *Die Augusteischen Handschriften 4. Cod. Guelf. 77.4 Aug. 2° – 34 Aug. 4°*, Frankfurt am Main: Klostermann, 1966.

H. Hermelink, Die ältesten magischen Quadrate höherer Ordnung und ihre Bildungsweise, *Sudhoffs Archiv für Geschichte der Medizin und der Naturwissenschaften*, XLII/3, 1958, pp. 199–217.

S. Heyne, *Die mittelalterlichen Handschriften der Universitätsbibliothek Marburg*, Wiesbaden: Harrassowitz, 2002.

I. Hiebner, *Mysterium Sigillorum, Herbarum & Lapidum*, Erfurt: In Verlegung Johann Birckners Buchhändlers, 1651.

Ibn Khaldūn, *The Muqaddimah: an Introduction to History*, 3 vols., transl. F. Rosenthal, New York: Pantheon Books, 1958.

Jordanus, an International Catalogue of Mediaeval Scientific Manuscripts, online via the *Ptolemæus Arabus et Latinus* project, Bavarian Academy of Sciences and Humanities, Alfons-Goppel-Straße 11, 80539 Munich, Germany, http://jordanus.badw.de/data.htm.

E. Jørgensen, *Catalogus Codicum Latinorum Medii Ævi Bibliothecæ Regiæ Hafniensis*, Copenhagen: Prostrat in Aedibus Gyldendalianis, 1923 – 1926.

D. Juste, Les Alchandreana primitifs: étude sur les plus anciens traités astrologiques latins d'origine arabe (Xe siècle), Leiden, Boston: Brill: 2007.

V. Karpenko, Two thousand years of numerical magic squares, *Endeavour* 18, 1994, pp. 147 – 53.

D. Karr, ed., *Liber Lunæ*, MS British Library, Sloane 3826, ff. 84r–100r, 1997 – 2010, accessible online at: http://www.digital-brilliance.com/contributed/Karr/Solomon/

al-Khwārizmī, Le calcul indien (Algorismus): histoire des textes, édition critique, traduction et commentaire des plus anciennes versions latines remaniées du XIIe siècle, ed. André Allard (préface de Roshdi Rashed), Paris; Namur: A. Blanchard; Société des études classiques, 1992.

D.A. King, *A survey of the scientific manuscripts in the Egyptian National Library* [= *American Research Center in Egypt (ARCE) Catalogs*, vol. 5], Winona Lake [Ind.]: Eisenbrauns, 1986.

Kircher, Athanasius, *Arithmologia, sive de abditis numerorum mysteriis* (*Pronicorum Propietatibus, Caput III, De Mathematico fabrica & constructionis dictorum Sigillorum ratiocinio*), Rome: Varesij, 1665.

Kircher, Athanasius, *Oedipus Aegyptiacus*, (III, *Caput* VI, *Tom.* I), Rome: Vitalis Mascardi, 1654.

F. Klaassen, Manuscripts of Astrological Image Magic, 2002, online at: http://homepage.usask.ca/~frk302/MSS/images.html.

P. Kunitzsch, The Transmission of Hindu-Arabic Numerals Reconsidered, in Jan P. Hogendijk and Abdelhamid I. Sabra, eds., *The Enterprise of Science in Islam: New Perspectives*, Cambridge: MIT Press, 2003, pp. 3 – 21.

B. Láng, *Unlocked Books: Manuscripts of Learned Magic in the Medieval Libraries of Central Europe*, University Park: Pennsylvania State University Press, 2008.

Late Medieval English Magic: English manuscripts containing 15th-century magical texts, online at: https://magicalmedieval.wordpress.com/2014/04/30/culddxi45/.

E. Leedham-Green, gen. ed., and J. Roberts, cons. ed., *Renaissance Man: the reconstructed libraries of European scholars, 1450 – 1700, Series one, The books & manuscripts of John Dee, 1527 – 1608, pt. 2. Manuscripts from Corpus Christi College, Oxford.* Marlborough, Wiltshire, England : Adam Matthew Publications, 1990-.

G. Leff, *Heresy in the Later Middle Ages, The Relation of Heterodoxy to Dissent c 1250–c.1450*, 2 vols., Manchester: Manchester University Press, 1967.

J.G. Lidaka, The Book of Angels, Rings, Characters and Images of the Planets: Attributed to Osbern Bokenham, in C. Fanger, ed., *Conjuring Spirits*, University Park: Pennsylvania State University Press, 1998, pp. 32 – 75.

P. Lory, La magie des lettres dans le *Shams al-maʿarif* d'al-Buni, *Bulletin d'études orientales* 39–40, 1987–1988, pp. 97–111.

P. Lory, Magie et religion dans l'oeuvre de Muhyî al-Dîn al-Bûnî, *Horizons Maghrébins*, 7/8, 1986, pp. 4 – 15.

Maslama b. Aḥmad al-Majrīṭī, *Picatrix: the Latin version of the Ghāyat al-ḥakīm: text, introduction, appendices, indices*, ed. David Pingree, London: Warburg Institute, University of London, 1986.

A. Marinoni, *Introduction to Luca Pacioli's De Viribus Quantitatis*, Milan: Enteraccolta vinciana, 1997.

Y. Marquet, Sabéens et Iḫwān al-Ṣafā', *Studia Islamica* 24, 1966, pp. 35–80; and 25, 1966, pp. 77–109.

M. Millàs, *Estudios sobre Azarquiel*, Madrid, Granada: C.S.I.C. Instituto Miguel Asin, 1943 –1950.

Manuel Moschopoulos, *Le Traité de Manuel Moschopoulos sur les carrés magiques, texte grec et traduction*, in *Mémoires scientifiques*, ed. and transl. P. Tannery, vol. IV, Paris & Toulouse: Jacques Gabay, 1912–1950), pp. 27–60 [= reprint of 1886 Bibliothèque Nationale ed.]

K.A. Nowotny, The construction of certain Seals and Characters in the work of *Agrippa* of Nettesheim, *Journal of the Warburg and Courtauld Institutes* 12, 1949, pp. 46–57.

Österreichische Nationalbibliothek (Handschriftenkatalog). *Tabulae codicum manu scriptorum, praeter graecos et orientales in Bibliotheca Palatina Vindobonensi asservatorum.* Edidit Academia Caesarea Vindobonensis. vol. IV: Cod. 5001–6500. Wien: Gerold, 1870.

Luca Pacioli, *De viribus quantitatis* (1498), MS Bologna, Biblioteca Universitaria di Bologna, cod. 250, part I, chap. 90.

S. Page, Magic at St Augustine's, Canterbury, in the Late Middle Ages, Phd diss., Warburg Institute, 2000.

W. Pagel, Review of *Paracelsus – Autor der Archidoxis Magica?* by W. Schneider, *Medical History* 27/3 (1983): 331.

D.A.M. Pielow, *Die Quellen der Weisheit: die arabische Magie im Spiegel des Uṣūl al-Ḥikma von Aḥmad ʿAli al-Būnī*, Hildesheim; New York: Georg Olms Verlag, 1995.

D. Pingree, The Diffusion of Arabic Magical Texts in Western Europe, in B. Scarcia Amoretti, ed., *La diffusione delle scienze islamiche nel medio evo europeo*, Rome: Accademia nazionale dei Lincei, 1987, pp. 57–102.

D. Pingree, Al-Ṭabari on the Prayers to the Planets, in A. Regourd and P. Lory, eds., *Sciences Occultes et Islam*, Damascus, 1993 (= *Bulletin d'Études Orientales* 44, 1992), pp. 105 –117.

Julius Reichelt, *Julii Reichelti Exercitatio, De Amuletis Aeneis figuris illustrata*, Argentorati: Apud J.F. Spoor & R. Wechtler, 1676.

C. Rieu, *Catalogue of the Persian Manuscripts in The British Library*, 3 vols., London: British Museum, 1879–1883.

A. Rodriguez, ed., Dos Fragmentos Inéditos de la Antigua Traducción Latina del De Plantis Duodecim Signis et Septem Planetis Subiectis Atribuido a Tésalo de Tralles, *Traditio* 59, 2004, pp. 369–82.

G. Rosińska, Scientific Writings and Astronomical Tables in Cracow: a Census of Manuscript Sources (XIVth-XVth Centuries), *Studia Copernicana* 22, Wrocław, 1984.

G. Saliba, The Role of the Astrologer in Medieval Islamic Society, in A. Regourd and P. Lory, eds., *Sciences Occultes et Islam*, Damascus: L'Institut Français d'Études Arabes de Damas, 1993 (= *Bulletin d'Études Orientales* 44, 1992), pp. 45–68.

W. Schneider, *Paracelsus, Autor der Archidoxis Magica?*, Bd. 23, *Veröffentlichung aus dem Pharmaziegeschichtlichen Seminar der Technischen Universität Braunschweig*, Braunschweig: [Technische Universität], 1982, pp. 145–170.

W. Schum, *Beschreibendes Verzeichnis der Amplonianischen Handschriften-Sammlung zu Erfurt.* Berlin: *Weidmannsche Buchhandlung*, 1887; available online at: http://www.manuscripta-mediaevalia.de/hs/kataloge/HSK0495.htm.

J. Sesiano, Construction of Magic Squares Using the Knight's Move in Islamic Mathematics, *Archive for History of Exact Sciences* 58, 2003, pp. 1–20.

J. Sesiano, Les carrés magiques de Manuel Moschopoulos, *Archive for History of Exact Sciences* 53, 1988, pp. 377–97.

J. Sesiano, *Les carrés magiques dans les pays islamiques*, Lausanne: Presses polytechniques et universitaires romandes, 2004.

J. Sesiano, Herstellungsverfahren magischer Quadrate aus islamischer Zeit (I), *Sudhoffs Archiv* 64, 1980, pp. 187–96.

J. Sesiano, Magic Squares for Daily Life, in Charles Burnett, Jan P. Hogendijk, Kim Plofker, and Michio Yano, eds., *Studies in the History of the Exact Sciences in Honour of David Pingree*, Leiden; Boston: Brill, 2004, pp. 715–734.

J. Sesiano, Quadratus Mirabilis, in Jan P. Hogendijk and A.I. Sabra, eds., *The enterprise of Science in Islam: New Perspectives*, Massachusetts: MIT Press, 2003, pp. 199–233.

J. Sesiano, Quelques constructions des carrés à magie simple dans les textes arabes, *Actes du 3me Colloque Maghrébin sur l'Histoire des Mathématiques Arabes* (Tipaza 1–3, Décembre 1990), Algérie: Association Algérienne d'Histoire des Mathématiques, 1998, pp. 251–62.

J. Sesiano, Quelques méthodes arabes de construction des carrés magiques impairs (I. Carrés a magie simple: a. Première methóde), *Bulletin de la Société Vaudoise des Sciences Naturelles* 83/1, 1994–1995, pp. 51–76.

J. Sesiano, Une compilation arabe du XIIe siècle sur quelques propriétés des nombres naturals, *SCIAMVS* 4, 2003, pp. 137–189.

J. Sesiano, *Un traité médiéval sur les carrés magiques, De l'arrangement harmonieux des nombres*, Lausanne: PPUR presses polytechniques, 1996.

M. Souissi, Ḥisāb al-wafq, *Cahiers de Tunisie* 30, HCT 16, 1978, pp. 27–43.

G.E. Szőnyi, The Occult Sciences in Early Modern Hungary in a Central European Context, in Blanka Szeghyová, ed., *The Role of Magic in the Past: Learned and Popular Magic, Popular Beliefs, and Diversity of Attitudes*, Bratislava: Pro-Historia, 2005, pp. 29–44.

P. Tannery, Les chiffres arabes dans les manuscrits grecs, *Revue archéologique* 3/7, 1886, pp. 355–60 [reprinted in tome IV, *Sciences Exactes Chez Les Byzantins*, of J.-L. Heiberg and H.-G. Zeuthen, eds., *Mémoires scientifiques*, 17 vols., Toulouse: Édouard Privat; Paris: Gauthier-Villars, 1912–1950, pp. 199–205].

Temple of the Sagrada Familia, Guide to the Passion of Christ Façade Sculptures by Subirachs, Barcelona: Editorial Mediterrània, 2001 (reprint).

L. Thorndike and P. Kibre, *A Catalogue of Incipits of Mediaeval Scientific Writings in Latin*, Cambridge: Medieval Academy of America, 1937.

A.J. Toynbee, *A Study of History*, 2nd ed., 12 vols., London; New York: Oxford University Press, 1935–61.

Carlos Fraenkel

On the Integration of Islamic and Jewish Thought: An Unknown Project Proposal by Shlomo Pines

The thesis about the dependency of medieval Jewish thinkers on their Arab intellectual milieu was most pointedly formulated by the great historian of philosophy and religion, Shlomo Pines, in a hitherto-unknown description of a project dating from 1959. Dr. Martin Ritter recently discovered this project sketch in a Harvard archive and entrusted it to me for publication. The project proposed by Pines is a history of "medieval Arabic and Jewish philosophy" from the ninth to the seventeenth century. According to Pines, "the integration of a history of Jewish philosophy into a history of Arabic philosophy is 'methodologically justified,'" because "in its decisive period mediaeval Jewish thought was an offshoot of Arabic thought: the debates of Jewish philosophers can only be understood" against the background of "the doctrinal differences obtaining among Arabic philosophers." There is a good dose of provocation in this thesis: the most creative period of Jewish thought, Pines argues, is nothing more than "an offshoot of Arabic thought"!

Pines's thesis is not only provocative, however, but also subversive for all those who believe in an unchanging essence of Judaism underlying its varying historical forms. What is this essence supposed to be if Jewish thinkers as significant as Saadya Gaon, Judah Halevi, Bahya ibn Paquda, or Maimonides, who laid the foundations of Jewish piety, theology, and philosophy in the Middle Ages, are inconceivable without their non-Jewish sources? To this day their works are studied in orthodox circles and thus continue to shape what it means to be Jewish.

To cite just one example of the questionability of conventional demarcations: important concepts of the Hasidic movement in Eastern European Judaism have their origins in Islamic Sufism – of course not directly, but through the mediation of Bahya ibn Paquda, whose thinking was strongly influenced by Sufism.[1] Bahya's main work, *Kitāb al-hidāya ilā farāʾiḍ al-qulūb* (Guide to the Duties of the Heart), describes the ascent of the soul to God. The Arabic original was translated into Hebrew in the twelfth century. The Hebrew translation, *Sefer hovot ha-levavot*, was one of the cornerstones of the genre of Hebrew ethical lit-

1 On Bahya's Sufism, see Lobel, *A Jewish-Sufi Dialogue*.

erature – the so-called *musar* literature – and remains a bestseller in religious circles, as well as an important source of Jewish piety.

As if that were not enough: *Toledot Yaakov*, the first published book of the Hasidic movement, authored by a student of the Baal Shem Tov in the second half of the eighteenth century, contrasts the insignificant "small struggle" several times with the more important "great struggle." The "small struggle" refers to a battle conducted with weapons; the "great struggle," however, refers to the moral wrestling of the soul with the "evil inclination" (*ha-yetzer ha-ra'*). This motif goes back to a famous *ḥadīth* that is frequently cited by Sufi mystics. It relates how the Prophet Muḥammad receives a group of soldiers with the words that they were returning from the "smaller *jihād*" – the *jihād* of the sword, to the "greater *jihād*" – the *jihād* of the soul "against pleasure." Can we therefore say that the Baal Shem Tov and his disciples studied the sayings attributed to the Prophet Muḥammad and their interpretation by Sufi mystics? Of course not! Their source is, as mentioned, Bahya ibn Paquda, who uses the *ḥadīṯ* – naturally without bibliographical details – in the fifth book of *Sefer hovot ha-levavot* for his own purposes.[2]

Unfortunately, Pines did not carry out the sketched project. It undoubtedly remains a desideratum.[3] To illustrate his basic thesis, a good example is Moses Maimonides (d. 1204), regarded as the most important Jewish philosopher of the Middle Ages. Many even consider him to be the most important Jewish philosopher of all time. On looking more carefully, however, it is not clear why we should characterize his work as "philosophical," or what exactly in his philosophy is supposed to be "Jewish." Indeed, strictly speaking, not a single work by Maimonides belongs to a genre of philosophical literature. If any, his short treatise, *Maqāla fī ṣināʿat al-manṭiq* (Treatise on the Art of Logic), likely written in his youth, could be described as a genuinely philosophical work. The work's authorship, however, has recently been disputed; and even if Maimonides is the author, the treatise falls into the category of ancient and medieval prolegomena literature – it is, as it were, on the threshold of philosophy, but not a philosophical work properly speaking.[4]

Like Averroes (Ibn Rushd, d. 1198), his famous Muslim colleague, Maimonides, too, is an outstanding commentator. But unlike Averroes, he does not

2 See Fenton, Judaeo-Arabic Mystical Writings, pp. 87 – 101.

3 For a good account of the methodological premises that guided Pines in his research on Jewish thought, see Harvey, Professor Pines, pp. 1 – 15.

4 On the problem of Maimonides's authorship, see Davidson, *Moses Maimonides*, pp. 313 – 22. On prolegomena literature in the Islamic world, see Jaffray, On the Threshold of Philosophy.

comment on the works of Aristotle, as we would expect a medieval philosopher to do, but on the Law of Moses.[5] His first important work, written in Arabic and entitled *Sirāj* (Light), is a commentary on the Mishnah. His last important work, customarily referred to as his philosophical magnum opus, is the *Dalālat al-ḥā'irīn* (Guide of the Perplexed), likewise written in Arabic. Maimonides himself does not present the *Guide* as a work of philosophy, but of biblical exegesis. According to the Introduction, his intention is to explain to perplexed Jewish intellectuals "the meaning of certain terms" and "of very obscure parables occurring in the books of the prophets."[6] And this, in order to show them that there are no genuine contradictions between the teachings of the prophets and the assertions of the philosophers.

What about the Jewish character of Maimonides's thought? First of all, it was perfectly natural for Maimonides to compose his works in Arabic, the language of philosophy and science in his day. But more important is what he demands in a famous passage of his already-mentioned commentary on the Mishnah: we ought to "listen to *al-ḥaqq* from whoever says it."[7] (*Al-ḥaqq* means "the truth" in Arabic; it is also one of God's names.) In the instructions that he sent to a student in response to the question which philosophical works are worth studying, he does not recommend a single Jewish author. Besides the Greeks, in particular Aristotle and his commentators, the philosophers he praises are all Muslims: al-Fārābī (d. ca. 950) who "excelled in wisdom," for example, or Ibn Bājja (d. 1138) whose "treatises are all good for the person who understands."[8] In another passage he expresses praise for Averroes's commentaries on Aristotle.[9] Of course he does not praise his Muslim colleagues because they are Muslims, but because he considers them good philosophers.

What becomes clear, then, is how much Maimonides, as a philosopher, felt at home in the tradition of Greco-Arabic philosophy and science. And he was right to be proud of belonging to this tradition.[10] From the eighth century to the tenth, excellent translations into Arabic were made of a large part of the corpus of Greek scientific and philosophical texts. It was an impressive achievement: one civilization appropriated the knowledge of another and turned it

5 In the Latin Middle Ages, Averroes was regarded as the commentator *par excellence* on Aristotle; see, e.g., Dante, *The Divine Comedy*, Inferno 4.144.

6 *Dalālat al-ḥā'irīn* (ed. S. Munk and I. Joel), p. 2; Eng. translation by S. Pines (*The Guide of the Perplexed*), pp. 5–6.

7 Maimonides, *Eight Chapters*, p. 154.

8 Maimonides, *Iggerot ha-Rambam*, p. 553.

9 Ibid., p. 299.

10 See Kraemer, Maimonides and the Spanish Aristotelian School, pp. 40–68.

into the basis of a vibrant intellectual culture of its own. This, moreover, was not the project of some isolated intellectuals; it was a large-scale enterprise carried out under the patronage of the political, social, and economic elite of the ʿAbbāsid caliphate (the second Sunni dynasty that ruled the Muslim empire after taking power from the Umayyads in 750). After the Greeks, the next significant period in the history of philosophy and science thus unfolded within Islamic civilization.[11] Its main intellectual centers were Baghdad (the residence of the ʿAbbāsid caliphs), and al-Andalus (Muslim Spain) – the last bastion of Umayyad rule. This is where Maimonides was born in the first half of the twelfth century. Like Averroes, Maimonides is a native of Córdoba, the seat of the caliphate of Córdoba, which in the tenth century was one of the most sophisticated cultural centres in Europe.

If we now bring together the two aspects of Maimonides's work that I briefly sketched above, I propose answering the question about Maimonides's philosophy and Judaism as follows: although he is first and foremost a commentator on the Law of Moses, his commentary is a *philosophical* commentary whose sources we find in Greco-Arabic philosophy and science. It is on the basis of these sources that Maimonides interprets Judaism as a *philosophical religion* – a religion founded, he argues, by philosopher-prophets – which not only forms the moral character of those who live according to its laws, but directs them to the intellectual love of God – to physics, the study of God's creation, and to metaphysics, the study of God himself.[12]

It was Shlomo Pines who wrote the definitive scholarly study of Maimonides's Greco-Arabic sources.[13] In this study Pines showed that it is indeed impossible to understand Maimonides without reference to the philosophical, scientific, and theological discussions going on around him in the Islamic world. This assessment, however, does not hold true only for Maimonides – it applies to all the great Jewish thinkers who lived in the Muslim cultural sphere. In numerous papers Pines uncovered surprising intersections between Islamic and Jewish intellectual history. Among his most interesting contributions is his analysis of the *Kuzari* by Judah Halevi (d. 1141), medieval Judaism's most important apologist. Pines succeeded in proving that central theological concepts for Halevi, which

11 On this process, see Gutas, *Greek Thought, Arabic Culture.*

12 I have discussed Maimonides's interpretation of Judaism as a philosophical religion in *From Maimonides to Samuel ibn Tibbon* (see in particular Chap. 2.2). For a comparison with a similar interpretation of Islam, see my article Philosophy and Exegesis, pp. 105 – 125.

13 See "Translator's Introduction: The Philosophical Sources of *The Guide of the Perplexed*" in Pines's English translation (pp. cviii – cxxiii).

had always been seen as authentically "Jewish" – for example, the concept of "chosenness" (*ṣafwa* in Arabic, *segulah* in Hebrew) – were in reality drawn from the Muslim Shīʿa, especially from the theology of the Ismāʿīliyya.[14]

Jewish thought flourished in the Muslim cultural sphere from the tenth to the twelfth century. Maimonides was the last great representative of this period – after him, the most significant developments in Jewish thought took place in the Jewish communities of Christian Europe. This, however, does not mean that Arabic philosophy lost its influence, for among the main sources for Jewish philosophers in Europe were Hebrew translations of many of the important works of Muslim and Judeo-Arabic philosophers, scientists, and theologians. These texts remained authoritative until the Early Modern period – and, in orthodox circles, remain so to this day – although, in the course of time, Christian thinkers, too, became increasingly influential.

Pines's project description contains not only a provocative – indeed subversive – thesis on the relationship between Jewish and Arabic thought. It also shows that his studies, like those on Maimonides and Halevi, should likely be seen as preliminary to a comprehensive and systematic work. We can only speculate about what prevented Pines from carrying out this project. What we do know is that, from the 1960s onwards, he became increasingly interested in the impact of Christian scholasticism on late medieval Jewish philosophy.[15] The resulting studies suggest that Pines would likely have modified the thesis underlying his 1959 project proposal in order to take the influence of Latin philosophy into account.

Finally, let us ask how persuasive Pines's thesis is. His claim that we cannot understand Jewish thought from the ninth to the seventeenth century without knowledge of its Arabic background is sufficiently substantiated in his published studies on Maimonides, Judah Halevi, and others. But what about his stronger thesis – that Jewish thought was only an "offshoot" of Arabic thought? I doubt that Pines himself would have consistently maintained the thesis in its most radical form. It is worthwhile to return once more to Maimonides, in many respects the paradigmatic example of the intertwined worlds of medieval Jewish and Arabic thought. In the course of his scholarly career Pines repeatedly examined Maimonides's work, highlighting aspects of his thought which clearly show that it was no mere "offshoot" of Arabic philosophy. Rather, Maimonides

---

14 Pines, Shīʿite Terms and Conceptions, pp. 165–251 (reprinted in *Studies in the History of Jewish Thought* [Collected Works, vol. 5], which volume contains numerous other papers illuminating the relationship between Jewish and Arabic thought).

15 Pathbreaking in this respect was his 1967 essay: Scholasticism after Thomas Aquinas and the Teachings of Hasdai Crescas and his Predecessors.

can be described as one of Arabic philosophy's most original representatives.[16] The view that Jewish thought, in order to be original, must be distinctly "Jewish" (or else it is no more than a pale copy of Arabic thought), is based on a false dichotomy. Jewish thinkers in the Islamic world – Saadya Gaon, Judah Halevi, Bahya ibn Paquda, Maimonides, and others – frequently stand out through the originality of their responses to the great philosophical, theological and scientific questions of their time – regardless of the fact that they were Jews who wrestled with these questions within a Jewish framework.

The title of the project sketched by Pines is *A History of Arabic and Jewish Mediaeval Philosophy.* He submitted the project to the Bollingen Foundation in New York as part of an application for a research grant. The description was forwarded for evaluation to Harry Austryn Wolfson (d. 1974), who had been a professor at Harvard University since 1925. Like Pines, Wolfson is among the most important twentieth-century historians of philosophy and religion. Although their scholarly interests overlapped in many ways, Pines and Wolfson usually reached quite different conclusions, which reflected in part the differences in their basic methodological assumptions.[17] No report by Wolfson on Pines's project proposal is extant. The proposal is located in the Harvard University Archives, *H.A. Wolfson Papers*, Sign. HUG (FP) 58.7, Box 30, folder "Pines," 1959.

A Detailed Statement of the Project

The history of the mediaeval Arabic and Jewish philosophy with which the present project is concerned will cover roughly the period from the ninth to the seventeenth centuries. It will deal with the transmission of Greek learning to the Islamic world, with the evolution of natural philosophy, metaphysics, and the political and sociological doctrines of the Arabic and Jewish philosophers, and to a certain extent with the impact made by these philosophers on the intellectual life of Mediaeval Christian Europe.

The main texts which will be studied are written in Arabic, Persian, and Hebrew. The integration of a history of Jewish philosophy into a history of Arabic philosophy is methodologically justified by the fact that in its decisive period mediaeval Jewish thought was an offshoot of Arabic thought: the debates of

16 Cf., e.g., Pines, The Limitations of Human Knowledge, pp. 82–109; and *idem*, Truth and Falsehood, pp. 95–157.
17 On Wolfson's approach, see Harvey, Hebraism and Western Philosophy, pp. 103–10. On Pines's approach, see n. 3, above.

the Jewish philosophers can only be understood if one refers to the doctrinal differences obtaining among Arabic philosophers.

Within the last thirty or forty years, our knowledge of Arabic thought has made great progress. In spite of this fact no comprehensive history of Arabic philosophy exists in any language. The short textbooks written on this subject are completely outdated. No history of Jewish philosophy on the lines suggested above has ever been written.

An adequate history of Arabic philosophy is essential for our understanding of the evolution of Islamic civilization. It should make a significant contribution to the understanding of the process of transmission and evolution of Greek learning in the Middle Ages, should render accessible to students doctrines of important Arabic and Jewish philosophers, who are barely known by name even to well-informed historians of philosophy, and should help to situate these [p.2] philosophers in the context of their times. Finally, it should help to bring about a clearer understanding of the Greek philosophical tradition which was modified by the Arabs and the Jews and transmitted in its new metamorphosis to Christian Europe. According to my present plans, the history of Arabic and Jewish philosophy will comprise two volumes of 500 to 600 pages each. Many important Arabic and Jewish philosophical texts have never been published and exist only in manuscript, or, if published, are very rare.

A comprehensive history of Arabic and Jewish philosophy can only be written if there is a possibility of working in various great libraries and collections of MSS in the Orient (in particular in Istanbul), in Europe, and this country. As the duration of an academic year in Jerusalem, where I teach, is only six to seven months, I could devote five to six months in the year to visiting libraries. The grant for which I apply would enable me to make in the course of three years the various journeys necessary for the project, to stay in each of the various countries for as long a time as is required by the search for and the study of important MSS, and for other forms of research required by the project.

I plan to finish writing the history within approximately three years from the beginning of my work on the project.

S. Pines

Bibliography

Dante, *The Divine Comedy*, Eng. trans. John Ciardi, New York, NY: New American Library, 2003.

H.A. Davidson, *Moses Maimonides: The Man and his Works*, Oxford: Oxford University Press, 2005.

P. Fenton, Judaeo-Arabic Mystical Writings of the XIIIth–XIVth Centuries, in Norman Golb, ed., *Judaeo-Arabic Studies*, Routledge, 1997, pp. 87–101.

C. Fraenkel, *From Maimonides to Samuel ibn Tibbon: The Transformation of the* Dalālat al-Ḥā'irīn *into the* Moreh ha-Nevukhim [Heb.], Jerusalem: The Hebrew University Magnes Press, 2007.

C. Fraenkel, Philosophy and Exegesis in al-Fārābī, Averroes, and Maimonides, *Laval Théologique et Philosophique* 64, 2008, pp. 105–125.

D. Gutas, *Greek Thought, Arabic Culture: The Graeco-Arabic Translation Movement in Baghdad and Early Abbasid Society*, London and New York: Routledge, 1998.

W.Z. Harvey, Hebraism and Western Philosophy in the Historiography of H.A. Wolfson [Heb.], *Daat* 4, 1980, pp. 103–10.

W.Z. Harvey, Professor Pines and his Approach to Jewish Thought [Heb.], *Jerusalem Studies in Jewish Thought* 7, 1988, pp. 1–15.

A. Jaffray, On the Threshold of Philosophy: A Study of Al-Fārābī's Introductory Works on Logic, PhD. Diss., Harvard University, 2000.

J. Kraemer, Maimonides and the Spanish Aristotelian School, in Mark D. Meyerson and Edward D. English, eds., *Christians, Muslims, and Jews in Medieval and Early Modern Spain: Interaction and Cultural Change*, Notre Dame: Notre Dame University Press, 1999, pp. 40–68.

D. Lobel, *A Jewish-Sufi Dialogue: Philosophy and Mysticism in Bahya ibn Paquda*, Philadelphia: University of Pennsylvania Press, 2006.

Maimonides, *Dalālat al-ḥā'irīn*, ed. S. Munk and I. Joel, Jerusalem: Junovitch, 1930–1931 [Eng. trans.: *The Guide of the Perplexed*, trans. S. Pines, Chicago: University of Chicago Press, 1963].

Maimonides, *Eight Chapters*, in J. Kafih, ed. and trans., *Mishnah 'im perush rabbenu Mosheh ben Maimon*, Jerusalem: Mossad ha-Rav Kook, 1963–1968.

Maimonides, *Iggerot ha-Rambam*, ed. I. Shailat, Jerusalem: Maaliyot Press, 1988.

S. Pines, The Limitations of Human Knowledge according to Al-Fārābī, Ibn Bājja, and Maimonides, in Isadore Twersky, ed., *Studies in Medieval Jewish History and Literature*, Cambridge, MA: Harvard University Press, 1979, pp. 82–109 [reprinted in S. Pines, *Studies in the History of Jewish Thought* (Collected Works, vol. 5), ed. Warren Zeev Harvey and Moshe Idel, Jerusalem: The Hebrew University Magnes Press, 1997, 404–431].

S. Pines, Scholasticism after Thomas Aquinas and the Teachings of Hasdai Crescas and his Predecessors [Heb.], *Proceedings of the Israel Academy of Sciences and Humanities* 1, 1967, pp. 1–101 [reprinted in S. Pines, *Studies in the History of Jewish Thought* (Collected Works, vol. 5), ed. Warren Zeev Harvey and Moshe Idel, Jerusalem: The Hebrew University Magnes Press, 1997, pp. 489–589].

S. Pines, Shī'ite Terms and Conceptions in Judah Halevi's *Kuzari*, *Jerusalem Studies in Arabic and Islam* 2, 1980, pp. 165–251 [reprinted in S. Pines, *Studies in the History of Jewish

Thought (Collected Works, vol. 5), ed. Warren Zeev Harvey and Moshe Idel, Jerusalem: The Hebrew University Magnes Press, 1997, pp. 219–305].

S. Pines, Truth and Falsehood Versus Good and Evil. A Study in Jewish and General Philosophy in Connection with the Guide of the Perplexed, I, 2, in Isadore Twersky, ed., *Studies in Maimonides*, Cambridge, MA: Harvard University Press, 1990, pp. 95–157.

Index

Arabic personal names are listed according to *kunya* (Abū al-...) or, if lacking, first name. Exceptions are for persons better known by some other designation. The Arabic particle *al-* is ignored for alphabetization. Greek, Hebrew and western European personal names are normally rendered with the forename in first position, unless there is a well established tradition of reference to the family name (e.g. Chaucer).

al-ʿAbbāsa, sister of caliph Hārūn al-Rashīd 53
ʿAbbāsid Caliphate 5–6, 26, 39–61, 103
ʿAbd al-Masīḥ 67
ʿAbd al-Raḥmān b. ʿUmar al-Ṣūfī 104
abjad numerals 166, 171
Abraham Ibn Daud 7, 80–81
Abraham b. ʿEzra 73–74
Abū al-ʿAbbās Aḥmad b. Muḥammad b. Kathīr al-Farghānī 6, 69, 104
Abū ʿAlī al-Khayyāṭ 70
Abū ʿAlī al-Manṣūr 70
Abū ʿAlī Yaḥyā b. Īsā b. ʿAlī b. Jazla 147–148
Abū Bakr 70
Abū Bakr Muḥammad b. Zakariyyāʾ al-Rāzī *see* Rhazes
Abū Isḥāq Ibrāhīm ibn Yaḥyā al-Naqqāsh al-Tujībī al-Zarqālluh 10, 92–93, 67, 106, 159–196
Abū Maʿshar Jaʿfar b. Muḥammad b. ʿUmar al-Balkhī 65, 70, 72
Abū al-Qāsim Khalaf al-Zahrāwī 137
Abū al-Rayḥān al-Bīrūnī 104
Abū al-Ṣalt Umayya al-Dānī 134–135
Achaia 9–10
Adelard of Bath 1–5
Aegidius of Tebaldis 70
agents and agency 2–11
agere (Latin verb) 3
Agrippa, Cornelius 11
Aḥmad b. ʿAlī al-Būnī 162–163
Aḥmad b. Yūsuf 66–67, 70
Alan of Lille 8, 124, 126–127, 130
Aleppo 70
Alexander of Ville-Dieu (Villa-Dei) 8, 124
Alexandria 44, 65
Alfanus, archbishop of Salerno 154
Alfonsine Tables 68, 71, 92
Alfonso X, King of Castile 10, 159–198
ʿAlī b. Aḥmad al-ʿImrānī 70
ʿAlī b. Abī al-Rijāl 70, 72
ʿAlī b. Khalaf 106–107
ʿAlī b. Riḍwān 70
ʿAlī b. Sahl Rabbān al-Ṭabarī 50
al-Andalus 42, 67, 99, 202
al-Andarzaghar 70
Aomar *see* ʿUmar b. al-Farrukhān
Aragon 6, 134
Aristotle 19, 46. 49, 67, 79, 83, 99, 102, 135, 201
Armengaud Blaise 8, 136–137
Armengaud of Prophatius 73
Arnau de Vilanova 8, 134–135
Arnulf of Orléans 121–122, 124
astrolabe 6–7, 67–69, 85–120
astrology 6, 63–76, 99–100
Astronomia Ypocratis 72
astronomy 6, 10, 15, 19, 23, 63–76, 86–120
Autolycus of Pitane 65–66
Avicenna *see* Ibn Sīnā
Azarquiel *see* Abū Isḥāq Ibrāhīm ibn Yaḥyā al-Naqqāsh al-Tujībī al-Zarqālluh

Baal Shem Tov 109
Bachelard, Gaston 108–109
Baghdad 26, 55, 70, 94, 97, 106, 202
Banū Mūsā 52
Barcelona 7, 71, 77–78, 135
– La Sagrada Familia (basilica) 160
Barmakid family 5–6, 56
Bartholomaeus Anglicus 8, 124–125
Barnard, Johannes 14

Bahya ibn Paquda 199–200, 204
Bayt al-Ḥikma 17
Benson, Larry D. 111
Berengar des Cortey 78
Bernard Silvester 8, 124–126, 130
Bernat de Berriac 8, 138
Bible 79–81
al-Bīrūnī *see* Abū al-Rayḥān al-Bīrūnī
al-Biṭrūjī *see* Nūr al-Dīn Abū Isḥāq al-Biṭrūjī
Blasgrave, John 10
Bollingen Foundation 202
Boniface VIII, Pope 136
Borrelli, Arianna 104
Bos, Gerrit 134–135
Buhktīshū' family *see also* Jūrjīs b. Jibrīl b. Bukhtīshū' 44, 52
Byzantine Empire 13–38, 44–45

Cairo 42, 70, 100
Capitula Almansoris 71
Cardano, Girolamo 11, 161, 170
Carmody, Francis J. 63
Castiglioni, Luigi 122
Catalonia 67, 77–84, 137–142
Centiloquium Bethen 72
Centiloquium Hermetis 72
Charles I of Anjou, King of Sicily 9–10, 145–157
Charles V, King of France 114
Chaucer, Geoffrey 7, 85–120
Clement V, Pope 136
climate (astronomical) 91
Comes, Mercè 159, 164
Constantine the African 14, 154
Constantinople 18, 22–38
– Monastery of Ioannes Prodromenos 27–28
Cornelius Agrippa 161, 165, 169, 190
cosmology 74
Corpus astronomicum 71, 73
Crescas, Hasdai 6–7, 77–84

Damascus 42, 99
Daniel, Pierre 192
Daremberg, Charles 14
David Caslari 134
Dāwūd b. Sarābiyūn al-Mutaṭannin 53
Dēnkard 46
Desiderius, abbot of Monte Cassino 154
Dioscordes 27
Dürer, Albrecht 160

Eagleton, Catherine 94–95, 113
Eisner, Sigmund 94
English (Middle) as language of science 87–89
ephemerides 91
Euclid 65–66, 105
Everard of Béthune 8, 124

al-Faḍl b. Sahl b. Zādhānfarrūkh 70
al-Fārābī 201
Farag (Faragius) of Agrigento (Faraj ibn Sālim) 9, 146–149
al-Farghānī *see* Abū al-'Abbās Aḥmad b. Muḥammad b. Kathīr al-Farghānī
Fāṭima Umm Muḥammad 53
fi'l 4
Firmicus Maternus 73
Frederick II, Emperor 150–151
al-Fullānī al-Kishnāwī 165

Gabrieli, Francesco 15
Galen 19, 134, 137
Galenism 5–6, 39–61
Gemma Frisius 106
geomancy 74
Georgios Chioniades 23 n. 58
Gerard of Cremona 6, 66–67, 69, 93, 133–134
Gerbert of Aurillac 104
Gergis 72, 74
al-Ghazālī 99–101
Giovanni da Capua 8, 136–137
Giovanni of Monte Cassino 9, 149
Guesdon, Marie Geneviève 17
Guillem Corretger 8, 137
Guillem Selva 141
Guarico, Luca 74
Gutas, Dimitri 41

al-Hādī 48
Ḥafṣid emirs of Tunis 153
al-Ḥākim 99

Hambly, Gavin 54
Hārūn al-Rashīd, caliph 53, 56
Henry Bate of Malines 73
Hermannus Contractus 105
Hervagius, Johannes 73
Horden, Peregrine 45
Hugo of Santalla 68, 71
humanism 77
Ḥunayn b. Isḥāq 49–50, 55, 66, 70, 135
Hypsicles 66

Ibn Abī Uṣaybiʿa 42
Ibn Bājja 98, 201
Ibn Daud, Abraham *see* Abraham ibn Daud
Ibn al-Haytham 68, 99–100, 165, 174
Ibn al-Jazzār, Abū Jaʿfar Aḥmad b. Ibrāhīm b. Abī Khālid 5, 14, 16, 19–22
Ibn Juljul 42
Ibn Khaldūn 162
Ibn al-Muthannā 68
Ibn Muʿādh 67
Ibn al-Nadīm 42, 50
Ibn Rushd 99–100, 200
Ibn al-Ṣaffār 67, 87, 93, 104
Ibn Shahlāfā 48
Ibn al-Shāṭir al-Dimashqī 99–100
Ibn Sīnā, Abū ʿAlī al-Ḥusayn b. ʿAbdallāh 24, 134–135, 150, 152
Ibn Ṭufayl 100
Ibn Yūnus 98
Ikhwān al-Ṣafāʾ 162
Immanuel ben Jacob Bonfils 73
instruments, astronomical 85–120
al-ʿImrānī *see* ʿAlī b. Aḥmad al-ʿImrānī
Ioannes Zacharias 24
ʿĪsā Abū Quraysh 50, 52–53
Ismāʿīliyya 201
Italy *see also* Sicily *and particular locations e.g.* Palermo, Salerno 20, 22

Jābir b. Aflaḥ 67, 100
Jacon ben David ben Yom Tov (Bonjorn) 73
Jacquart, Danielle 135
Jean de Nesle 9, 149
Jesus 82
Jewish scholars 77–84,
Jewish translators 134, 140–143, 146–149
Jibrīl b. Bukhtīshūʿ 50–53
Joan I, King of Aragon 7, 77–78, 82–83
Johannes Hispaliensis *see* John of Seville
Johannes Regiomontanus 70
John Damascene 23 n. 55
John, Duke of Bedford (d. 1435) 192
John of Pavia 68
John of Sacrobosco 68–69, 87, 105
John of Saxony 71
John of Seville 6, 69, 94, 104, 163
John of Stendhal 71
John Philoponus 103
John Somer 93, 112
Jordanus 64
Judah Halevi 199, 202–204
Jundishapur 17, 44
Jūrjīs b. Jibrīl b. Bukhtīshūʿ 43–44, 48

Kairouan 70
Khayzurān, wife of caliph al-Mahdī 52–53, 55
Khosraw Anūshirwān, Shah 47
al-Khwārizmī *see* Muḥammad b. Mūsā al-Khwārizmī
Kibre, Pearl 64
al-Kindī, Abū Yūsuf 52, 70, 73–74
King, David A. 96, 103–104
Kircher, Athanasius 11, 161, 171
Kittridge, George Lyman 111
Konstantinos Meliteniotes 23
Kouzis, Aristoteles 15

Laird, Edgar 112
Lanfranc of Milan 9, 140–141
Lindberg, David 115
London 73
Louis IX, King of France 153

madrasa 97
magic squares 10, 159–196
al-Mahdī, caliph 49
Mallorca 139–142
Maimonides, Moses (Ibn Maymūn) 8, 11, 135–137, 199–204
Manfred, King of Sicily 150
Manilius 73
Manuel Moscopoulos 161, 165, 168, 190

al-Manṣūr, Abū Yūsuf Yaʿqūb b. Yūsuf b. ʿAbd al-Muʾmin, caliph 41, 43–44, 47–48, 103
Manuscripta Mediaevalia 64
manuscripts
– Bologna, Biblioteca universitaria 3632 23 n. 58
– Cairo, Dār al-Kutub Sh124 164
– Cairo, Dār al-Kutub TJ 424 165–166, 174
– Cambridge, University Library Add. 4087 177
– Cambridge, University Library Dd. XI 45 181
– Copenhagen, Kongelige Bibliotek GKS 1685 179
– Erfurt, Wissenschaftliche Bibliothek CA 4° 361 178–179
– Erfurt, Wissenschaftliche Bibliothek CA Math. 54 191
– Florence, Biblioteca Medicea Laurenziana Antinori 101 35
– Frankfurt, Universitätsbibliothek lat. oct. 231 168
– Ghent, Universiteitsbibliotheek Hs 1021 186–187
– Graz, Universitätsbibliothek 342 138–143
– Krakow, Biblioteka Jagiellońska BJ 793 182
– Leiden, Bibliotheek der Rijksuniversiteit Voss. gr. F 65 14 n 10
– London, British Library Add. 9599 165, 172
– London, British Library Harley 2404 175
– London, British Library Persian Add. 7713 188–189, 191
– London, British Library Sloane 3826 186
– London, Wellcome Library 60 29
– London, Wellcome Library 128 185
– Marburg, Universitätsbibliothek Mscr. 18 183
– Munich, Bayerische Staatsbibliothek CLM 4610, 14482, 14809 122 n. 9
– Oxford, Bodleian Library Baroccianus 150 29
– Oxford, Bodleian Library Holkham Greek 109 184–185
– Oxford, Corpus Christi College 125 176
– Paris, Bibliothèque nationale de France, esp. 212 138–142
– Paris, Bibliothèque nationale de France, gr. 2194 29, 35
– Paris, Bibliothèque nationale de France, gr. 2219 29
– Paris, Bibliothèque nationale de France, gr. 2260 29
– Paris, Bibliothèque nationale de France, gr. 2286 27
– Paris, Bibliothèque nationale de France, gr. 2287 29
– Paris, Bibliothèque nationale de France, gr. 2309 29
– Paris, Bibliothèque nationale de France, lat. 6912 9, 146, 149-
– Vatican City, Biblioteca Apostolica Vaticana, Ottoboni lat. 1809 177–178
– Vatican City, Biblioteca Apostolica Vaticana Pal. gr. 279 29
– Vatican City, Biblioteca Apostolica Vaticana Pal. lat. 1407 73
– Vatican City, Biblioteca Apostolica Vaticana Reg. lat. 1283a 167
– Vatican City, Biblioteca Apostolica Vaticana Vat. gr. 300 20, 22
– Vatican City, Biblioteca Apostolica vaticana Vat. lat. 2398–2399 10, 149
– Vienna, Österreichische Nationalbibliothek 162d (= Arabic Flügel 1421) 172–174
– Vienna, Österreichische Nationalbibliothek 1421 164, 173
– Vienna, Österreichische Nationalbibliothek lat. 5239 180
– Vienna, Österreichische Nationalbibliothek med. gr. 21 29
– Wolfenbüttel, Herzog August Bibliothek Guelf 17–8 Aug. 184
Maragha 97, 99
Māsawayh family 52
Māshāʾallāh 69–70, 72, 87, 93, 104
Maslama al-Majrīṭī 67, 87, 93
mathematics 159–196
Matthew of Vendôme 124
McGill University Research Group on Transmission, Translation and Transformation in Medieval Textual Cultures 1–2

medicine *see also* Galenism
- Arab-Islamic 13–38, 39–61
- Byzantine 13–38, 44–45
- Indian 5, 50
Menelaus of Alexandria 66
Mercati, Giovanni 14
Messiah
Messina, monastery of San Salvatore 21–22
Metge, Bernat 6–7, 77–84
Michael VII Doukas, Emperor 20
Michael Psellos 20
Michael Scot 67
Minardus Theutonicus 149
Mogenet, Joseph 15
Morrison, J.E. 105–106
Montpellier 135–136, 138
Moses of Palermo 147
Muḥammed, Prophet 190
Muḥammad ibn Ibrahīm al-Fazārī 103
Muḥammad b. Mūsā al-Khwārizmī 68–105
al-Mustanṣir, caliph 70
Mu'ayyad al-Dīn al-'Urḍī 100

Naples 146, 151
Naṣīr al-Dīn al-Ṭūsī 100, 104
navicula (instrument) 106
Neophytos Prodromenos 27
Nicholas of Lynn 93, 112
Nisibis 44
Normandy 73
Nūr al-Dīn Abū Isḥāq al-Biṭrūjī 67, 99–100
Nutton, Vivian 17, 44

observatories 97–100
Orléans 122
Osborne, Marijane 87, 90
Ovid, *Metamorphoses:* Vulgate Commentary on 7–8, 121–132

Pacioli, Luca 161, 183–184
Palermo 21–22
Pamphilus de amore 124
Pappus of Alexandria 65–66
Paul the Persian 46
Pèlerin de Prusse 113–114
Pentalogos, Gerasimos 15
Pere de Queralt 78
Perpignan 73
Persia *see also* Sassanid dynasty, Shāpūr, Khosraw Anūshirwān 23, 45–47, 79, 104
Pethion, Mar 49
Petrus Philomena 68
philosophy
- Islamic 199–205
- Jewish 11, 199–205
physicians as translators 13–38, 39–61, 133–143
physics 74
Picatrix 193
Pierre Bersuire 121
Pines, Shlomo 11, 199–207
Pisa 73
pharmacy 25–27
Plato of Tivoli 70–71
Pormann, Peter 44
Prophatius Judaeus (Jacob Machir ibn Tibbon) 69, 73
Ptolemy 6, 65–66, 69–71, 99, 102

al-Qabīṣī, Abū al-Ṣaqr 'Abd al-'Azīz 71–72
quadrant 69
al-Qifṭi 42–43, 55
Quṭb al-Dīn al-Shīrāzī 100

Ragep, Jamil 16, 95
Reggio di Calabria 20
Regius, Raphael 121
Rhazes (al-Rāzī) 9–10, 18 n. 30, 19–20, 23 n. 55, 145–157
Rheinberger, Hans 108
Ritter, Martin 199
Robert the Englishman 69
Robert the Wise, King of Sicily 155
Romanos II Porphyrogenitus, Emperor 18 n. 33
Rome 136

Saadya Gaon 199, 204
Sabians 162
Sabra, A.I. 97, 99, 114
al-Shabrāmallisī 165
Sacrobosco *see* John of Sacrobosco

Ṣadr al-Sharīʿa al-Thānī 100
Shaghab, mother of caliph al-Muqtadir 55–56
Sahl b. Bishr 70, 72
Salerno 146–147, 151, 154
Saliba, George 100–101, 115
Salio of Padua 71, 74
Samarkand 98
Sarābiyūn family 52
Saragossa 77–78, 83
Sassanid dynasty 41, 44–47
Savage-Smith, Emilie 44
Sayf al-Dawla 70
Sayılı, Aydın 97
Sayyid Rukn al-Dīn 98
Schism, Great (of the Church) 82–83
Seville 67
Sezgin, Fuat 64
Shādhān b. Baḥr 64, 70
Shāpūr, Shah 45
Shatzmiller, Joseph 134
Shefer, Miri 55
Sicily 5, 9–10, 18, 145–157
soul, human 78–81
Spain *see also* al-Andalus, Aragon, Catalonia 6, 66–68, 152
square, magic *see* magic square
Steinschneider, Moritz 141
surgery 137–143
Symeon Seth 4, 19–20, 23 n. 55

al-Thaʿālibī 45
al-Ṭabarī *see* ʿAlī b. Sahl Rabbān al-Ṭabarī
Tarascon, Tables of 73
Tayfūrī family 52
Teodorico Borgognoni 8, 137-
textual cultures, medieval 1–11
Thābit b. Qurra 66, 69
Theodosius, mathematician and astronomer 66
Thorndike, Lynn 64
Tihon, Anne 15
time-reckoning 90, 96–98, 107
Timothy the Patriarch 49–50, 53
Toledo 66–67, 71, 106
– Tables of 68, 92
Toledot Yaakov 200
Tours 8
translation and translators 13–38, 39–61, 63–76, 133–143, 145–157
– Arabic to Castilian 10
– Arabic to Greek 4–5, 13–38
– Arabic to Latin 6, 8, 9, 63–76, 105, 134, 136–137, 145–157
– Catalan to Hebrew 9
– French to English 87
– Greek to Arabic 39–61, 134
– Greek to Latin 6, 63–76, 150
– Greek to Pahlavi 44, 46
– Greek to Syriac 44
– Hebrew to Catalan 7, 78
– Hebrew to Latin 6, 73
– Italian to English 87
– Latin to Catalan 137–143
– Latin to English 87–120
– Latin to Hebrew 134, 140–142
– Persian to Arabic 50
– Persian to Greek 13–38
– Sanskrit to Arabic 50, 103
– Syriac to Arabic 49
Tudela 73
Tunisia 10, 145, 153–154
Ṭūsī couple 99

Ulugh Beg 98
ʿUmar b. al-Farrukhān 70, 74

Valencia 134–135
Van Arsdall, Anne 45
Violant (Yolande), Queen of Aragon 7, 77–78, 83

Walter of Châtillon 8, 124–130
Wolfson, Harry Austryn 204
women as patrons of translation 53–55

Yūḥannā b. Māsawayh 50

al-Zahrāwī *see* Abū al-Qāsim Khalaf al-Zahrāwī
al-Zarqālī *see* Abū Isḥāq Ibrāhīm ibn Yaḥyā al-Naqqāsh al-Tujībī al-Zarqālluh
Zubayda, wife of caliph Hārūn al-Rashīd 53–54

www.ingramcontent.com/pod-product-compliance
Lightning Source LLC
Chambersburg PA
CBHW051541230426
43669CB00015B/2676